Lectures on Phase Field

 **MyCopy**

Ingo Steinbach • Hesham Salama

Lectures on Phase Field

 Springer

Ingo Steinbach
Ruhr University Bochum
Bochum, Germany

Hesham Salama
Ruhr University Bochum
Bochum, Germany

https://doi.org/10.1007/978-3-031-21171-3

This work was supported by Ruhr-Universität Bochum

Cover figure: Dendrites in MnAl solidification with melt flow ©RUB, Tegeler.
Portrait: Prof. Steinbach ©RUB, Marquard.

This Springer imprint is published by the registered company Springer Nature Switzerland AG
The registered company address is: Gewerbestrasse 11, 6330 Cham, Switzerland

Preface

This book is, more or less, the script of a lecture series entitled "Phase-Field Theory and Application," which I give at Ruhr University Bochum (RUB) in my department Scale-bridging Thermodynamic and Kinetic Simulation (STKS). It strongly rests on my personal experience during many years in academic and applied research with a special focus on "phase field." The idea is to combine a very brief presentation of the materials-related mechanisms alongside the theoretical background of the phase-field theory and, last but not least, to introduce numerical solutions. The lecture series is designed for graduate students and newcomers to phase-field theory. There are many aspects of phase field that will only be covered briefly in this script. At my university, we are continuing this lecture series with an advanced course in the format of a discussion seminar. We hope that we can continue with this in a second volume of the book.

The 12-lecture course, as held at RUB, is condensed to seven lectures in this book. An eighth lecture, "Quantum Phase Field," is added here that I do only in the advanced course: Just free your mind and enjoy!

Exercises are recommended at the end of each lecture, and suggestions for further reading are given. Examples are added where appropriate. A short introduction to the phase-field code OpenPhase academic [1] is added as a "tutorial" with two examples as a beginning to your own research.

Bochum, Germany   Ingo Steinbach
June 2022

Reference

1. Openphase, https://openphase.rub.de/

Contents

Part I
The Phase-Field Method

Chapter 1
Introduction

1.1 What Is a Phase Field?

A phase field, in the context of these lectures, is a scalar field in space and time that indicates the local state of matter as characterized by its phase state. This may not help much for now: do not worry, we will proceed step by step, explaining all expressions. Firstly, the present section will explain what a phase field is—i.e., what do we understand by this expression? Secondly, in Sect. 1.2, we will investigate which types of materials problems can be examined using phase fields. In Sect. 1.3, the historical background of phase field will be detailed; perhaps you will come back to this part at the end of the lectures, at which time you should be better able to reference the different historical aspects of phase field comparing to their current state of development. In the last section of this introductory chapter, Sect. 1.4, we will focus in particular on the scale of the materials problem: the mesoscopic scale. Let us begin!

The "phase" of matter, a lump of atoms within a reference volume, denotes the state of crystalline order between the atoms. It is a central concept in physics and a central element of the phase-field theory. The order of atoms—i.e., their position in a crystalline lattice: FCC, BCC, or other crystalline structures—defines the phase state. In addition, the amorphous, liquid, and gaseous states—i.e., the absence of crystalline order—are characterized as phases. If the phase state is a gas, we may be referring to an atmosphere outside of a solid-state sample or a pore inside the sample. Liquids are mainly considered as a melted phase in connection with a solid crystal. This can be water and ice, or molten iron in connection with already-solidified iron and slag. Further types of phase are characterized by magnetic or ferroic order, superconductivity, or plasma. The phase field, a field in space and time, indicates the state of a materials point (in space and time), whether solid, liquid, gas, plasma ferromagnetic, etc, It tells nothing about other properties of matter at this point (in space and time), like its temperature, pressure, the composition of elements or molecules. Of course this information is also needed. But it is not enough to

© The Author(s) 2023
I. Steinbach, H. Salama, *Lectures on Phase Field*,
https://doi.org/10.1007/978-3-031-21171-3_1

characterize the material at this point (...). We need to know it's phase state, which is enclosed in the local value of the "phase field."

In the physics literature, the phase state is characterized by a so-called "order parameter," which is normalized between 0 and 1. We will use the letter ϕ to represent this parameter. The term "order" relates to "crystalline order," which is different between different phases. For example, a crystalline solid will be stable at low temperatures, while its melt will be stable above the melting temperature. An important element of the concept of phases is that there is a discontinuity in their order; the order does not change gradually if, for example, the temperature changes: there is an abrupt change in the atomic order (at least for phase transformations of the first kind, which we will be considering here). This discontinuity is also reflected in discontinuities in the properties of the phases, such as their elasticity, viscosity, thermal or electrical conductivity. In other words: the phase state can be determined uniquely from the material properties of the piece of matter we are investigating.

Another important aspect is that the phase state needs not to be stable. We also consider metastable or unstable phases. This means that two materials with the same temperature, pressure and composition may be in different phase states. In other words, to characterize the state of a material not only pressure, temperature and composition is needed, but also information about the atomic order: its phases state.

To make this concept more transparent to materials scientists, let us discuss a metallic alloy, say a binary aluminium–silicon alloy, regarding possible phases at different temperatures and compositions. Figure 1.1b shows the phase diagram of this alloy. This is called a "eutectic" phase diagram, as it has three stable phases in different regimes of alloy composition c and temperature T, as indicated in the figure: liquid, FCC aluminum, and FCC silicon (diamond lattice) on the far right-hand side of the diagram is almost pure silicon, which is not shown. These regimes in which a phase is stable are also termed "phase fields" in alloy thermodynamics. Don't be confused. They are fields in temperature, composition, and pressure, not in space and time as we will use the term in the context of these lectures. But both usages of "phase field", of course, are somewhat related.

In-between these stable regions of individual phases, we have the so-called "two-phase" regions, where none of the adjacent phases can be stable. The reason for this can be read from the Gibbs energies of the system as function of composition [see Fig. 1.1a]. For a given temperature and pressure, the Gibbs energy functions G_α and G_β of the two phases, α and β, are displayed schematically. The minimum of the total Gibbs energy is determined by the so-called double-tangent construction, where the common tangent touches the curves at the phase compositions c_α and c_β. With the fraction of phases $f_\alpha = 1 - f_\beta$, we define the total Gibbs energy G_{total}:

$$G_{\text{total}} = f_\alpha G_\alpha + f_\beta G_\beta. \tag{1.1}$$

The total Gibbs energy is minimal along the tangent line between c_α and c_β, weighted with the fractions f_α and f_β. Each individual phase with composition c ($c_\alpha < c < c_\beta$) will have a higher energy. Therefore, the material with a nominal composition c_0 in the two-phase region must decompose in two phases

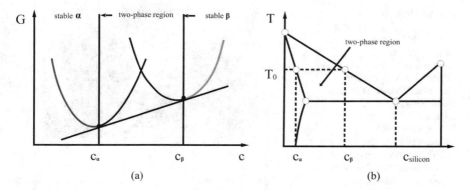

Fig. 1.1 (a) Schematic of the Gibbs energy curves of a binary alloy in two different phase states, α and β, for a given temperature T_0. The common tangent determines: (i) the phase compositions c_α and c_β of an alloy with nominal composition within the two-phase region, and (ii) the minimum Gibbs energy of a two-phase mixture with this nominal composition. (b) Linearized Al–Si phase diagram, indicating the stable composition of c_α and c_β for the given temperature T_0

so that the material will reach a lower state in the Gibbs total energy. We say that the composition region between c_α and c_β is "forbidden" for both phases. There is an "energy barrier" in the Gibbs energy of the material that separates it into different phases. Other properties of the different phases—such as elasticity, thermal conductivity, and density—also show a similar behavior, and they are clearly different in both phases.

All of this is well known to most of our readers, but we repeat it here to perform the following thought experiment. The upper panel of Fig. 1.2 displays a cross section through a two-phase mixture between liquid (white phase) and FCC aluminum (black phase) close to equilibrium. A line scan of the local composition is performed, and the corresponding composition $c(x)$ is displayed in the lower panel of this figure. This switches from c_α to c_β and back again. It is now an easy exercise to normalize this composition to determine the phase fractions as a function of space, as also displayed along the line scan:

$$f_\alpha = \frac{c - c_\beta}{c_\alpha - c_\beta}, \qquad f_\beta = \frac{c - c_\alpha}{c_\beta - c_\alpha}; \qquad f_\beta = 1 - f_\alpha. \tag{1.2}$$

The fractions f_α and f_β are usually used to characterize a phase mixture as a whole; the fractions are not considered as fields in space and time. Therefore, we use a different notation: we use $\phi_\alpha(x)$, the "phase field," which varies in space between 0 and 1. Note that we will also consider interfaces between phases as "diffuse," i.e., with a finite width. Therefore, the graph in Fig. 1.2 is not step-wise: it is "smeared out," as will be discussed in detail in Sect. 1.3.1. The normalized function "phase field" is also called an "indicator function" because it indicates where a phase is present in space. In the same way as it is defined from the varying phase concentrations in (1.2), it could be defined from the elasticity or the density of

Fig. 1.2 Scheme of a
solid–liquid phase mixture
(solid in black, liquid in
white) close to the
equilibrium of the
compositions within the
individual phases. The scan
evaluates the local
composition, which varies
between the solid
concentration c_α and the
liquid concentration c_β. On
the left-hand axis, the value
of the phase field ϕ_{solid} is
displayed, with β indexing
the solid phase

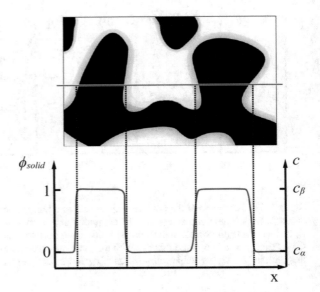

the individual phases. We can also map the phase field back to the local composition,
elasticity, and density if we know the local phase-field value and the properties of
the individual phases.

The phase field is a field in space that indicates the position at which we find a
special phase α, $\phi_\alpha(x) = 1$, or do not find this phase, $\phi_\alpha(x) = 0$. Then we let this
evolve in time (!):

$$f_\alpha \rightarrow \phi_\alpha(x) \rightarrow \phi_\alpha(x, t).$$

We can see that a multi-phase material can be characterized regarding its local
phase state. We can use the standard concepts of alloy thermodynamics to determine
the local phase fractions in a small reference volume containing, say, a thousand
atoms to be large enough to characterize the phase state reliably. Then, we will find
either the values of 1 or 0, or if our reference volume is intersected by an interface,
the value will be *between* 1 and 0. Interfaces may move, and phases can change.

The content of the following chapters involves establishing how to determine the
evolution of phases in complex materials under various conditions.

1.2   What Is the Purpose of Phase Field?

Generally speaking, the phase-field method uses a set of partial differential equations to describe a material's problem with evolving microstructures. Let us first
elaborate on what we mean by "evolving microstructures." The microstructure of a

multicrystalline material—and we will mainly speak about the microstructures of metallic materials—is the distribution and shape of different grains in that material. These individual grains have many attributes: their orientation with respect to a reference orientation; their composition, pressure, and temperature; along with important crystalline defects such as dislocations, stacking faults, and vacancies. If two neighboring crystalline grains are of the same phase but have different orientations, they form a "grain boundary." Consequently, they are considered as different elements of the microstructure and will be denoted by different indicators, that is, by different phase fields. Two grains of different phases form a "phase boundary," and a pore within a solid grain forms a "surface."

We will describe all these different cases using phase fields. This means that a phase field marks a region of space in which the material can be i) characterized uniquely by its phase state and orientation or ii) as an interface (where the phase state is undetermined). Practically speaking, the phase field $\phi_\alpha(\vec{x})$, a field in three-dimensional (3D) space $\vec{x}$, has, by convention, the value $\phi_\alpha(\vec{x}) = 1$ for the material point at position $\vec{x}$ belonging to the phase α indexed by ϕ_α. For $1 > \phi_\alpha(\vec{x}) > 0$ it is an interface, for $\phi_\alpha(\vec{x}) = 0$ it is another phase than α. Other conventions can be found in the literature, but this is the prevailing one. If there are only two grains in our material, e.g., a small inclusion in a matrix, we will skip the subscript α and index the inclusion by $\phi(x) = 1$ and the matrix by $\phi(x) = 0$ (or vice versa). The phase boundary is marked by an intermediate value $0 < \phi(x) < 1$. Treating the field as continuous, or "diffuse," as we do in phase-field theory, the boundary has to have a finite width η.

Such a construction is very helpful for characterizing a polycrystal, a multi-phase material, or any general static microstructure. We want, however, to allow the microstructure to change in time, $\phi(x) \rightarrow \phi(x, t)$. Now we have to distinguish two cases as follows. (i) The microstructure is just "deforming," i.e., we find a mapping of space coordinates $\vec{x}(t_1) \rightarrow \vec{x}(t_2)$. In continuum mechanics, this is called the mapping from the "reference frame" at time t_1 to the "actual frame" at time t_2. In the second case, (ii) the microstructure is "evolving"; it may be growing, swelling, or shrinking. It may have been born in the time $t_1 \rightarrow t_2$ ("nucleated," as we say in materials science), or it may have died, i.e., it has fully transformed to a different phase. Thus, it cannot just be referenced in two different frames. In general, one material point in an evolving structure transforms to a material point of a different kind (e.g., liquid to solid). No physical movement of matter has to be involved in this process. We will speak in these chapters about case (ii), evolving microstructures. We will in addition allow them to deform (see Chap. 7), but this is a side remark.

The example shown at the end of this chapter is a dendrite growing in two dimensions (2D). In this example, which we undertake as an exercise, only the interface between the phases (solid and liquid) "moves." This means that a material point, which had been liquid in the starting condition, changes its phase state to solid. However, no material is moving, neglecting the effects of shrinkage during the phase transformation and motion of the solid and the melt. For small changes in the shape of the dendrite, we could also treat this problem using finite elements. Then, the elements would have to be swelling or shrinking in both phases. However, in the general case, if we start from full liquid and simulate to full solid, we will need so-called adaptive finite elements, which continuously adapt to the shape of the dendrite. This is not impossible: it has been demonstrated by several authors, particularly in 2D simulations, but also in 3D (e.g., by Provatas et al. [16] and Schmidt [17]). However, this is numerically expensive, particularly in 3D. The phase-field method offers an elegant and numerically efficient way to treat this problem of evolving, growing, or shrinking microstructures. Aside from using adaptive finite elements, alternative approaches include the cellular automata and level set methods (see Further Reading).

Having described a materials problem—e.g., solidification or coarsening of a microstructure during heat treatment—as a moving-boundary problem using a set of partial differential equations, we will also need external boundary conditions and initial conditions. Furthermore, we will need a good database for physical input data and constitutive models for all the properties of the bulk phases and boundaries. Last but not least, we will need good solvers and sound numerics. All of this will be touched on during these lectures. It should be noted that the scientist applying the method will have to be patient: a good numerical solution to a problem may take weeks or months. ...

1.3   History of the Phase-Field Method

1.3.1   Microscopic Phase Field

The phase-field method can be split into two branches with very different histories, interpretations, and intentions. The first branch is often referred to as "order-parameter theory," the "microscopic phase-field method," or the "time-dependent Ginzburg–Landau theory." This rests on grounds in thermodynamics with a special focus regarding interfaces. Commonly, van der Waals is noted as being the father of this branch; he rationalized from general considerations that a diffuse interface between two phases will be more likely than a sharp interface [23], although at this time, the existence of the atomistic structure of matter had not yet been established. As a next step, Landau introduced the concept of an order parameter into the thermodynamic description of materials. Ginzburg must be referenced as introducing gradients of the order parameter into the concept, representing interfaces or phase boundaries. Other stepping-stones in the development are the Cahn–Hilliard theory of spinodal decomposition [3] and Khachaturyan's theory of microelasticity [8]. Wheeler, Boettinger, and McFadden introduced a first model of alloy solidification in 1992 [24], and this combines a Cahn–Hilliard model of a material with a miscibility gap and a phase-field model of the second type, which we will call a "mesoscopic phase-field model" below. A compilation of both branches with notable contributions, by far incomplete, is shown in Fig. 1.3. Before giving the historical details of the second branch—mesoscopic models—let us provide an

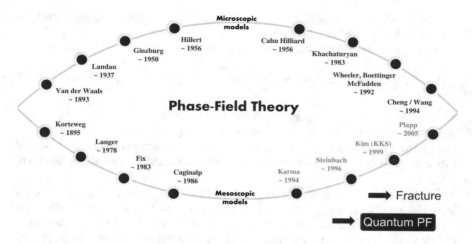

Fig. 1.3 Two branches of phase-field theory—microscopic and mesoscopic—highlighting important steps in their development. Today, the branches converge to a common understanding accepting the strengths of each approach. We stop this history about 20 years ago, so newer developments in phase-field theory are not included, such as models of fracture [6, 14, 18] or quantum phase fields [11, 21]

outline of the problem that urges us to extend microscopic models, along with its possible solution.

1.3.2 The Problem of Scale

All phase-field models—and this must be very clear—agree in terms of their general structure; they start from a thermodynamic functional density $f(\phi, \nabla\phi, T, c, \epsilon, \ldots)$ with three terms of different character:

$$f = \frac{\epsilon}{2}(\nabla\phi)^2 + \frac{\gamma}{4}\phi^2(1-\phi)^2 + m(\phi, T, c, \epsilon, \ldots). \tag{1.3}$$

Here, we use the notation of Kobayashi [9]. The first and second terms relate to interface or grain-boundary contributions. This is easy to see, since the first term vanishes in the bulk when $\phi = const$, i.e., in the bulk of the grains $\phi \equiv 1$ or $\phi \equiv 0$ (in the convention we shall use throughout this script). For these conditions, the second term—the so-called "double-well potential"—also vanishes by construction. Both terms represent a positive energy penalty within the interface, $0 < \phi < 1$, which is related to the interface energy (see Chap. 2). The last term in (1.3) relates to the bulk energy difference between phases. This is a function of temperature T, composition c, strain ϵ, and other material states such as magnetism. Charges may be added, but those cases are not treated in this lecture series, and they are little investigated in phase-field theory.

We can also write down the functional in the form:

$$f = \frac{\sigma}{\eta}\left[\eta^2(\nabla\phi)^2 + 72\phi^2(1-\phi)^2\right] + \Delta g(\phi, T, c, \epsilon, \ldots), \tag{1.4}$$

where σ is the interfacial energy, η is the interfacial width, and Δg is the energy difference between phases, the function m in Eq. (1.3) above. Now, if all the parameters of the equations—ϵ, γ, m, σ, η and Δg—are constants, we can find a direct one-to-one correlation between the two representations, as will be shown in Chap. 2.

We can take for granted that the relation between m and ΔG works, since it is not phase field specific: it is standard thermodynamic Gibbs energy difference between different phases. The relations between ϵ, γ, σ, and η have the following problem: if ϵ and γ are determined from an underlying microscopic theory, as well as the interface energy σ, the interface width η is uniquely fixed. If all parameters are correct, we will find, as in real materials, that the interface width $\eta \approx 1$ nm. This is good if we are investigating materials at the nanometer scale (as scientists do with microscopic phase-field models); it is not so good if we want to investigate microstructures at the micrometer scale, because it will not be feasible to treat the diffuse interface of a phase field with a width of 1 nm in a 3D simulation of micrometer-sized grain structures. How can this dilemma be solved? We seek a theory that can be matched to a so-called "sharp interface" theory (see Chap. 4). In this theory, the interface width has no physical importance. To match a diffuse-interface method (the phase-field method) to the theory of such a sharp interface, we need to remove the influence of the interface width on all quantities with physical meaning. We seek a theory that is agnostic regarding the interface width of a real material: the interface width shall be scalable for convenience of numerical simulation. To say this clearly: it is frustrating to throw away the physical insights of a phase field to predict the structure of an interface; it is, however, the great success of mesoscopic phase-field models to make quantitative predictions of microstructural processes at larger scales possible!

1.4   Mesoscopic Phase-Field Model

The mesoscopic phase-field model[1] stems from a very different route than thermodynamics. It is rooted in wave mechanics, in particular "traveling-wave solutions" called "solitons" in the physics literature. The investigation of these phenomena dates back to Korteweg, a Dutch mathematician in the nineteenth century. It is

---

[1] The phrase "phase field" was adopted by Jim Langer in 1978 [13] for a special phase transformation: solidification.

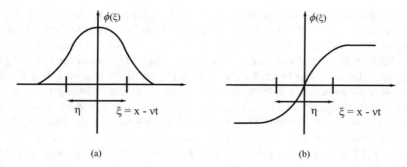

Fig. 1.4 Schematic representation of solitonian waves. The left-hand panel displays a wave pulse $\phi(\xi)$ of width η along the space coordinate x traveling with speed v; this remains self-similar in the moving coordinate $\xi = x - vt$. The right-hand panel shows a wave front; this is an integral form of the wave pulse. The width η and moving coordinate ξ are the same as for the wave pulse

reported that he observed a single wave packet, caused by a boat clutching against a channel wall in Amsterdam, his hometown. The wave travelled for a long distance without changing its shape. He and his collaborators wrote down a non-linear wave equation for such a phenomenon in the shallow waters of rivers or ocean shores. This is known today as the Korteweg–de Vries (KDV) equation; it is of third order in the space derivative, and its solution is displayed in Fig. 1.4a.

The special feature of this solution is self-similarity in its local coordinate $\xi = x - vt$ with the velocity v of the wave packet, withstanding moderate outer perturbations. This property, as realized by Langer, can be used to propagate interfaces by solitonian waves in a numerical simulation: the shape of the wave front is not part of the solution, i.e., it can be eliminated from the solution. The wave packet, Fig. 1.4a, hereby does not distinguish between either side of the packet, right or left. Therefore, in phase-field models, the integral form of the KDV equation is used as a second-order equation with an asymmetric solution, as displayed in Fig. 1.4b. This is more or less exactly what we call today phase field: the minimum solution of (1.3) or (1.4).

In the physics literature, both the symmetric and the anti-symmetric solutions are termed "solitons": self-similar traveling-wave solutions of non-linear wave equations, in contrast to periodic solutions of linear wave equations. There are further aspects of this connection between linear and non-linear wave equations that relate to the fundamental understanding of matter (see Chap. 8). In particular, soliton solutions can be used for quantization if applied to an appropriate wave function, as already pointed out by Olsen in 1974 [15]. This "quantization of phase field" has been applied to studying scale formation in the universe, [22]. Enthusiastically speaking, quantization of phase field opens the road to a new understanding of the physical world: space, time, and matter are unified in a wave-mechanical description that is consistent with both quantum mechanics and thermodynamics (see further reading, "Solitons and quantum phase field," on this topic).

Returning to the classical phase field in materials science, the idea is to combine the equation for the solitonian front, which travels with a given velocity v, with a transport equation in the bulk phases ahead or and behind the wave. The transport in the bulk phases ahead and behind, of course, is influenced on the one hand by the wave itself. On the other hand, the state of the phases ahead and behind will affect the wave velocity wave v. For the classical soliton, the front velocity is a constant such that the traveling wave is a plane wave; however, now in our applications, the velocity is a vector in 3D space, varying at each position of the wave front dependent on the interaction of the wave front with its surroundings: $v \rightarrow \vec{v}(\vec{x}, t)$.[2] The equation used to couple both phenomena is the empirical Gibbs–Thomson equation. This relates the velocity of the front linearly to the kinetic undercooling of a front with mobility M^ϕ and the front normal $\vec{n}$:

$$\vec{v} = \vec{n} \, M^\phi \left(\sigma^* \kappa + \Delta g \right). \tag{1.5}$$

The Gibbs–Thomson equation considers capillarity with the interface stiffness σ^* and the local curvature of the front κ; i.e., the kinetic undercooling is expressed as the curvature-corrected Gibbs energy difference between the bulk phases Δg. The Gibbs–Thomson equation is not dependent on the interface width η: it represents a sharp-interface model at the mesoscopic scale, where the interface can be approximated as a discontinuity between the phases. The Gibbs–Thomson equation (1.5) can be replaced by an appropriate phase-field equation (see Chap. 2). Since the width of the phase-field function in the transition between phases has no analogue in the Gibbs–Thomson equation, it can be chosen (within certain bounds) for numerical convenience. This will be treated in detail in Chap. 4.

It was also realized early on [1, 2] that the curvature κ is considered inherently in a phase-field description of a curved interface. This will be discussed in detail in Chap. 3.

1.4.1 Applications

The phase-field method now has many applications, from biophysics to astrophysics. Its main field of application is materials science, and metallurgy in particular. As described above, we distinguish the two main approaches: microscopic models, which are applied mostly to solid-state transformations (the school of Armen Khachaturyan); and mesoscopic models, which were first applied to solidification (the school of Jim Langer). Today, there are well-established methods for combining both branches for specific materials problems; these range, in

---

[2] Throughout these chapters, we will only use vector notation where it is needed for understanding.

the context of metallurgy, from solidification and heat treatment to in-service degradation until failure. A new branch has arisen with phase-field models of fracture [7, 14, 18]. This will, however, not be covered in this lecture series. One common aspect of all these applications is the evolution of the microstructure. Evolving structures define a special class of moving-boundary problems, in which phase-field models offer elegant and efficient solutions.

Example—Dendritic Growth
Dendritic solidification occurs if a solid crystal grows from an undercooled melt. As a result, the interface becomes unstable, forming tips and branches. It then develops into a "tree-like" structure: hence the term "dendrite."

Equiaxed Dendrite
This example is related to Kobayashi's dendrite in 2D [9, 10], and it is an exercise that we undertake in the class. One very good result from a former student is displayed in Fig. 1.5. The "100-line code" (see Appendix A.1), written in C++, evolves the phase-field variable and temperature field with adiabatic boundary conditions. One circular grain is placed at the bottom of the 2D simulation zone at the start of the simulation, and the system size is 300×600 grid points. The phase field ϕ is displayed for different time steps.

Further Reading
- Long-range elastic, electrostatic, and magnetic interactions connected to domain structure evolution during structural, ferroelectric, and ferromagnetic phase transformations, a recent review: [4].
- Fracture: [6, 14, 18].
- Level set: [5, 20].
- Hydrodynamic interpretation of Phase Field: [12]
- Solitons and quantum phase field: [11, 19, 21].

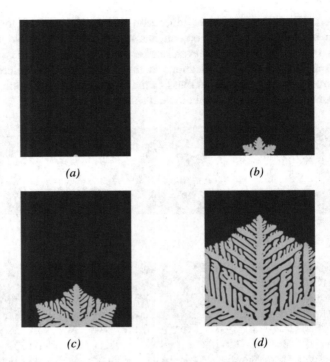

Fig. 1.5 Dendritic solidification of a pure substance into an undercooled melt. (**a**) $t = 0.05$ s. (**b**) $t = 0.25$ s. (**c**) $t = 0.75$ s. (**d**) $t = 1.5$ s

References

1. G. Caginalp, E. Socolovsky, Phase field computations of single-needle crystals, crystal growth, and motion by mean curvature. SIAM J. Sci. Comput. **15** (1994). https://doi.org/10.1137/0915007
2. G. Caginalp, W. Xie, Mathematical models of phase boundaries in alloys: phase field and Sharp interface, in *Motion by Mean Curvature and Related Topics: Proceedings of the International Conference held at Trento, Italy, 20–24, 1992*, ed. by G. Buttazzo, A. Visintin. (De Gruyter, 2011), pp. 43–62. https://doi.org/10.1515/9783110870473.43
3. J.E. Cahn, J.E. Hilliard, Free energy of a nonuniform system. I. Interfacial free energy. J. Chem. Phys. **28**, 258–267 (1958)
4. L.-Q. Chen, Y. Zhao, From classical thermodynamics to phase-field method. Progress Mater. Sci. **124**, 100868 (2022). ISSN:0079-6425. https://doi.org/10.1016/j.pmatsci.2021.100868
5. F. Gibou, R. Fedkiw, S. Osher, A review of level-set methods and some recent applications. J. Comput. Phys. **353**, 82–109 (2018). ISSN:0021-9991. https://doi.org/10.1016/j.jcp.2017.10.006
6. A. Karma, Phase-field formulation for quantitative modeling of alloy solidification. Phys. Rev. Lett. **87**(11), 115701 (2001)
7. A. Karma, D.A. Kessler, H. Levine, Phase-field model of mode III dynamic fracture. Phys. Rev. Lett. **87**(4), 045501 (2001). https://doi.org/10.1103/physrevlett.87.045501
8. A.G. Khachaturyan, *Theory of Structural Transformations in Solids* (Wiley, New York, 1983)

9. R. Kobayashi, Modelling and numerical simulations of dendritic crystal growth. Physica D **63**, 410–423 (1993)
10. R. Kobayashi, A numerical approach to three-dimensional dendritic solidification. Exp. Math. **3**, 59–81 (1994)
11. J. Kundin, I. Steinbach, Quantum-phase-field: from the Broglie–Bohm doublesolution program to doublon networks. Z. Naturforsch. **75**(2a) (2020). https://doi.org/10.1515/zna-2019-0343
12. A.G. Lamorgese, D. Molin, R. Mauri, Phase field approach to multiphase flow modeling. Milan J. Math. **79**, 597–642 (2011). https://doi.org/10.1007/s00032-011-0171-6
13. J.S. Langer, Unpublished research notes. Int. Mater. Rev. **64**(6) (1978). See appendix in W. Kurz, D. J. Fisher, R. Trivedi (2019) Progress in modelling solidification microstructures in metals and alloys: Dendrites and cells from 1700 to 2000, pages 311–354.
14. C. Miehe, F. Welschinger, M. Hofacker, Thermodynamically consistent phasefield models of fracture: variational principles and multi-field FE implementations. Int. J. Numer. Methods Eng. **83**(10), 1273–1311 (2010). https://doi.org/10.1002/nme.2861
15. S.O. Olsen, A natural way of quantization. Acta Physiol. Acad. Sci. Hung. **37**, 97–103 (1974)
16. N. Provatas, N. Goldenfeld, J. Dantzig, Efficient computation of dendritic microstructures using adaptive mesh refinement. Phys. Rev. Lett. **80**, 3308–3311 (1998). https://doi.org/10.1103/PhysRevLett.80.3308
17. A. Schmidt, Computation of three dimensional dendrites with finite elements. J. Comput. Phys. **125**, 293–312 (1996)
18. D. Schneider et al., Phase-field modeling of crack propagation in multiphase systems. Comput. Methods Appl. Mech. Eng. **312**, 186–195 (2016). https://doi.org/10.1016/j.cma.2016.04.009
19. A.C. Scott, F.Y.F. Chu, D.W. McLauchlin, The soliton: a new concept in applied science. Proc. IEEE **61**, 1443–1483 (1973)
20. O. Stanley, F. Roland, *Level Set Method and Dynamic Implicit Surfaces* (Springer, Berlin, 2002)
21. I. Steinbach, Quantum-phase-field concept of matter: emergent gravity in the dynamic universe. Z. Naturforsch. A **72**(1) (2017). https://doi.org/10.1515/zna-2016-0270
22. I. Steinbach, J. Kundin, F. Varnik, Self similarity of the expanding universe as understood by quantum-phase-fields (2020). arXiv: 2002.12848 [physics.gen-ph]
23. J.D. van derWaals, The thermodynamic theory of capillarity under the hypothesis of a continuous variation of density. J. Stat. Phys. (1979). Amsterdam (1893). Trans. J. R. Robinson, pages 197–244.
24. A.A. Wheeler, W.J. Boettinger, G.B. McFadden, Phase-field model for isothermal phase transitions in binary alloys. Phys. Rev. A **45**, 7424–7439 (1992)

Chapter 2
Analytics

2.1 The Problem of Propagating a Wave Front on a Numerical Grid

In this chapter, we will recapture the analytical background of traveling-wave solutions in 1D with two of the most commonly used separating potentials: the so-called "double-well" and "double-obstacle" potentials. This solution, also known as the antisymmetric soliton solution (see previous Chap. 1), is the basic reason that phase fields work so well in a numerical solution: the solution is quite robust against perturbations, which are inevitable in numerical simulations. Of course this resembles the physical fact, that the wave front is robust against any perturbation—wind, a bird etc.—in reality. The front self-stabilizes while traveling over the numerical grid, just like a solitonian wave on the surface of water.

Let us start with an example. We define a wave front $\phi(x, t_0)$ at time t_0 in one dimension x:

$$\phi(x, t_0) = \frac{1}{2} \left(1 - \tanh\left(\frac{3x}{\eta}\right) \right). \tag{2.1}$$

We know the wave front from Fig. 1.4b: the solitonian wave, but now defined between 0 and 1, as we do nowadays in phase field. In the following we take this convention that the left-hand side of the wave is $\phi = 1$ and its right-hand side $\phi = 0$. The front is centered around $x = 0$, and the width η is a measure of the distance between $\phi = 0.05$ and $\phi = 0.95$. This "cutoff" is necessary since the hyperbolic tangent converges to 0 or 1 only at infinity, and we need a finite measure of the interface width.

We now want to propagate this wave with constant speed v_0. One writes:

$$\dot{\phi}(x, t) = \frac{\partial \phi}{\partial x} \frac{\partial x}{\partial t} = \frac{\partial \phi}{\partial x} v_0. \tag{2.2}$$

I. Steinbach, H. Salama, *Lectures on Phase Field*,
https://doi.org/10.1007/978-3-031-21171-3_2

You may try to solve this equation numerically in 1D (as we do in the class), applying the profile (2.1) as a starting condition. However, this will simply not work for propagating the wave for a reasonable distance while maintaining its profile! Equation (2.2) is called a hyperbolic transport equation, and it is prone to large numerical instability up to shock-waves, and we do not want this at all. The equation thus has to be regularized using an appropriate method. We will not go into the details of numerical regularization approaches in hydro-dynamical wave mechanics, but we will modify the equation for better solvability. We want to solve (2.2) exactly. The only things we can do to the equation are: (i) add 0, or (ii) multiply by 1. We will do the first. We add:

$$0 = \frac{\partial^2 \phi}{\partial x^2} - \frac{72}{\eta^2} \phi^2 (1 - \phi)^2 \left(\frac{1}{2} - \phi \right). \tag{2.3}$$

Does this look strange? Let us do the derivation. We will prove that (2.3) has the solution (2.1) and that it is self-similar in time for a propagating planar wave, a wave in 1D. Therefore, it adds 0 to the original Eq. (2.2). We add it:

$$\dot{\phi}(x, t) = 0 + \frac{\partial \phi}{\partial x} v_0,$$

$$= \frac{\partial^2 \phi}{\partial x^2} - \frac{72}{\eta^2} \phi^2 (1 - \phi)^2 \left(\frac{1}{2} - \phi \right) + \frac{\partial \phi}{\partial x} v_0. \tag{2.4}$$

This is the famous "phase-field equation" we will be dealing with in the rest of the lectures.[1] Adding this strange 0 Eqs. (2.3)–(2.1) changes the type of the equation from hyperbolic to parabolic. It becomes a "diffusion equation," which is easy to solve in a discretized manner on a numerical grid. Let us go through this step by step.

2.2   Equation of Motion for the Phase Field

We start with the phase-field functional density, Eq. (1.3). A functional, by its mathematical definition, is a mapping of an arbitrary set of functions (functional densities), within a given definition domain, onto the real numbers. These functional densities, functions in space and time, are very difficult to compare. If we map functions onto a scalar, the comparison becomes easy. The functions can then be ordered on a linear coordinate indicating whether they are "larger or smaller." For a human, sadly speaking, "money" is an often-used measure; this measures the value of all goods that can be bought. In metallurgy, we like to use the free energy of the

---

[1] The equation is given here without physical prefactors, which will be used to multiply the 0 later. Why? To make you curious for Chap. 3: because in 3D, the "0" gets a physical meaning!

system because it is a thermodynamic functional of the state variables temperature and pressure, two quantities that are easily controllable in lab experiments. Other functionals are useful under different conditions. In thermodynamics, we use Legendre transformation to relate these functionals uniquely. Nevertheless, different functionals should be used for different applications.

To introduce "time," we start from an entropy functional S. The second law of thermodynamics demands that entropy increases with time:

$$0 < \frac{dS}{dt} = \frac{\partial \phi}{\partial t} \frac{\delta S}{\delta \phi}. \tag{2.5}$$

Here, the symbol δ denotes the "variational derivative" of a functional, i.e., of a function of functions. This will be elaborated below in some detail. For a rigorous derivation, see Kiselev, Shnir, and Tregubovich's book [2]. For now, let us simply take the message that Eq. (2.5) can easily be fulfilled if

$$\frac{\partial \phi}{\partial t} \propto \frac{\delta S}{\delta \phi} \quad \text{or} \quad \frac{\partial \phi}{\partial t} \propto -\frac{\delta F}{\delta \phi}. \tag{2.6}$$

This relation is termed the "Clausius–Duhem relation" in continuum mechanics. It is not a "physical law," but it is the simplest possible Ansatz to ensure the positivity of entropy production; it works well and is generally accepted. It is termed the "relaxation Ansatz" because it has only a first derivative in time and no accelerations. Since the entropy S enters the Gibbs free energy F with a "−" sign, we accept that:

$$\frac{\delta S}{\delta \phi} = -\frac{\delta F}{\delta \phi}.$$

$F = \int_\Omega d^3x f$ is the total free energy within the 3D domain Ω with the free-energy density f. We have to work on the functional as an integral over the domain of interest since the free energy density f of a phase-field model is a function of gradients in ϕ, $\nabla\phi$, which are, by construction, not defined on a special point in space but need at least two different points. Let us treat the three parts of the free energy in (1.3) separately:

$$f_1 = \frac{\epsilon}{2}(\nabla\phi)^2, \tag{2.7}$$

$$f_2 = \frac{\gamma}{4}\phi^2(1-\phi)^2, \tag{2.8}$$

$$f_3 = m(\phi, T, c, \epsilon, \ldots) = h(\phi)\Delta g(T, c, \epsilon, \ldots) = \frac{3\phi^2 - 2\phi^2}{6}\Delta g(T, c, \epsilon, \ldots). \tag{2.9}$$

We start with the easiest one: f_2. We are not afraid of the phase-field function $\phi(x, t)$; we treat it as a normal variable at the point in space x and time t

under consideration, and it should not depend on the values of the field at other locations regarding the function f_2. This will come later. Therefore, we can treat the variational derivative as a normal partial derivative. We calculate:[2]

$$\frac{\delta F_2}{\delta \phi} = \frac{\partial f_2}{\partial \phi} = \gamma \phi (1 - \phi) \left(\frac{1}{2} - \phi \right). \tag{2.10}$$

As mentioned before, this part is only active in the interface. It changes its sign at $\phi = \frac{1}{2}$, i.e., in the center of the interface. It is the first derivative of the so-called double-well potential, also called the Landau potential, $\phi^2 (1 - \phi)^2$. Therefore, we call it the potential contribution.

In the third term, f_3 (2.9), we have already separated the phase-field-dependent part $h(\phi) = \frac{3\phi^2 - 2\phi^2}{6}$, the so-called coupling function, from the physical part, i.e., the Gibbs-energy-density difference between the phases as a function of temperature, composition, strain, and other parameters, $\Delta g(T, c, \epsilon, \ldots)$. Here, we will take this as a constant $\Delta g = \Delta g_0$. Because this part also does not depend on non-local contributions, i.e., gradients of the phase field, the functional derivative can again be treated as a normal partial derivative:

$$\frac{\delta F_3}{\delta \phi} = \frac{\partial f_3}{\partial \phi} = \phi (1 - \phi) \Delta g_0. \tag{2.11}$$

As we can see, the special form of (2.9) has been chosen such that the free-energy difference acts only within the interface when $0 < \phi < 1$. Δg_0 is called the "driving force," since it makes a positive or negative contribution to the evolution of the phase field $\dot{\phi}$, driving it to grow or to shrink.

The first part (2.7) is more involved because it contains a gradient in ϕ, $(\nabla \phi)^2$, in 1D, $(\frac{\partial \phi}{\partial x})^2$. We will perform the variational differentiation in 3D so as not to increase the confusion of the problem with the dimensionality of the variational derivative. The 1D case works identically. Applying the chain rule, we find:

$$\frac{\delta F_1}{\delta \phi} = \frac{1}{\delta \phi} \delta \left(\int_\Omega d^3 x \frac{\epsilon}{2} (\nabla \phi)^2 \right) = \frac{1}{\delta \phi} \int_\Omega d^3 x \left(\epsilon \nabla \phi \delta (\nabla \phi) \right). \tag{2.12}$$

The difficulty lies in finding a way to treat the variation $\delta (\nabla \phi)$ of the gradient operator. It helps to realize that all derivatives—regardless of whether they are functional derivatives $\delta \phi$, partial derivatives $\partial \phi$, or partial derivatives in space

---

[2] Note that the common convention in physics and mathematics literature about variational derivatives δ violates the otherwise accepted convention that a symbol $\frac{\delta}{\delta}$ should be dimensionless. F is an extensive (absolute) quantity, while f is an intensive quantity (defined per volume). The variational derivative removes the volume integral and the volume increment $d^3 x$.

$\nabla\phi$—are linear operators and are therefore commutable. What we will do is commute the operators δ and ∇ by a partial integration in space:

$$\int_{\Omega} d^3x \left(\epsilon \nabla\phi \delta(\nabla\phi)\right) = -\int_{\Omega} d^3x \left(\epsilon \nabla^2\phi\right)\delta\phi + \text{boundary integral}, \quad (2.13)$$

$$\frac{\delta F_1}{\delta\phi} = -\epsilon\nabla^2\phi. \quad (2.14)$$

The boundary integral is omitted mostly with the reasoning that the interface should not touch the boundary of the domain. This will not be the case in most applications, so boundary conditions should always be treated with care. This topic, however, will only be touched on in some numerical exercises. The most important factor of the derivation is the change of sign in the gradient contribution (2.12) due to the partial integration. In the free-energy functional, both interface contributions (2.7) and (2.8) are positive penalties against forming interfaces; they will counteract with opposite signs in the equation of motion: The Laplace- or diffusion-operator will act as to smear out the profile, while the potential operator will act as to sharpen it, as we will see in detail. Collecting all terms defines the equation of motion of the phase field, the so-called phase-field Eq. (2.4), and it now has physical constants in Kobayashi's notation. It is derived from the Clausius–Duhem relation (2.6) with the proportionality constant τ, which is called the relaxation constant.

$$\tau \frac{\partial\phi}{\partial t} = \epsilon\nabla^2\phi - \gamma\phi(1-\phi)\left(\frac{1}{2} - \phi\right) - \phi(1-\phi)\Delta g_0. \quad (2.15)$$

For completeness, a common formal replacement of the functional derivative for theories with gradient contributions (quantum mechanics in general) has the following form, known as the "Euler–Lagrange equation," in which $\nabla\phi$ is treated as a symbolic entity in the denominator of the first differential operator:

$$\frac{\delta}{\delta\phi}F = \left\{-\nabla\frac{\partial}{\partial\nabla\phi} + \frac{\partial}{\partial\phi}\right\}f.$$

2.3   Traveling-Wave Solution for the Double-Well Potential

A "traveling-wave solution," as introduced in Chap. 1 (Fig. 1.4), has the special feature of self-similarity in time (for constant driving force Δg_0 and velocity v_0); i.e., we can define a coordinate $\xi = x - v_0 t$ to describe this solution in its own frame moving with constant velocity with regard to a resting frame. We derive this solution as the minimum solution of the functionals (1.3) or (1.4). Of course, the condition of self similarity $\frac{d\xi}{dt} = 0$ and the minimum energy condition are related, because a minimum solution requires the vanishing of the first derivative of the phase-field $\frac{\delta\phi}{\delta t} \propto \frac{\delta F}{\delta\phi} = 0$ in Eq. (2.15). Here, we will not try to derive the solution by applying some scheme to solve partial differential equations, but we will prove that the given solution from the literature does the job![3] The solution has already been introduced in (2.1) for the initial time step $t = 0$. We change $x \rightarrow \xi = x - v_0 t$ to find the traveling solution:

$$\phi(x, t) = \frac{1}{2} \left(1 - \tanh\left(\frac{3(x - vt)}{\eta} \right) \right). \tag{2.16}$$

To prove that this is a solution of (2.15), we have to compute the first and second derivatives in the space coordinate x:

$$\frac{\partial \phi}{\partial x} = -\frac{6}{\eta} \phi (1 - \phi), \tag{2.17}$$

$$\frac{(\partial \phi)^2}{\partial x^2} = \frac{72}{\eta^2} \phi (1 - \phi)(\frac{1}{2} - \phi). \tag{2.18}$$

We find:

$$\epsilon \nabla^2 \phi - \gamma \phi (1 - \phi)(\frac{1}{2} - \phi) = \left(\frac{72}{\eta^2} \epsilon - \gamma \right) \phi (1 - \phi) \left(\frac{1}{2} - \phi \right)$$

$$= 0 \quad \text{for} \quad \frac{72}{\eta^2} \epsilon = \gamma. \tag{2.19}$$

This is the first important result, and all phase-field models agree: the minimum solution of ϕ for a phase-field functional (1.3) with $\Delta g = 0$, or with a self-similar solution and arbitrary but constant velocity v, demands an interface width

$$\eta = \sqrt{\frac{72\epsilon}{\gamma}}. \tag{2.20}$$

Inserting this relation back into the phase-field Eq. (2.15), only the driving force part remains, and we have:

$$\tau \frac{\partial \phi}{\partial t} = -\phi (1 - \phi) \Delta g_0 = \frac{\eta}{6} \frac{\partial \phi}{\partial x} \Delta g_0. \tag{2.21}$$

We compare this with the Gibbs–Thomson equation for a moving planar interface, i.e., where the capillarity term is 0, and the interface mobility is M^ϕ:

$$v_0 = M^\phi \Delta g_0, \tag{2.22}$$

$$\frac{\partial \phi}{\partial t} = M^\phi \frac{\partial \phi}{\partial x} \Delta g_0. \tag{2.23}$$

---

[3] A reference to the first scientist who derived this solution is difficult. It is not a particular solution for a phase field; it is very general.

This relates the relaxation constant τ to the interface mobility M^ϕ:

$$\tau = \frac{\eta}{6M^\phi}. \tag{2.24}$$

Finally, we have to relate ϵ and γ to the interface energy σ. The interface energy (in units $\left[\frac{J}{m^2}\right]$) must come out if we integrate the free energy density (in units $\left[\frac{J}{m^3}\right]$) in the normal direction through the interface:

$$\sigma = \int_{-\infty}^{+\infty} dx \left[\frac{\epsilon}{2}(\nabla\phi)^2 + \frac{\gamma}{4}\phi^2(1-\phi)^2\right]$$

$$= \int_{-\infty}^{+\infty} dx \left(\frac{18}{\eta^2}\epsilon + \frac{\gamma}{4}\right)\phi^2(1-\phi)^2$$

$$= \int_{-\infty}^{+\infty} dx \quad \frac{1}{2}\gamma \quad \phi^2(1-\phi)^2$$

$$= -\int_0^1 d\phi \frac{\partial x}{\partial\phi} \quad \frac{1}{2}\gamma \quad \phi^2(1-\phi)^2 \tag{2.25}$$

$$= +\int_0^1 d\phi \quad \frac{1}{12}\eta\gamma \quad \phi(1-\phi) \tag{2.26}$$

$$= \quad \frac{1}{72}\eta\gamma \quad = \quad \frac{\epsilon}{\eta}. \tag{2.27}$$

We have used the first derivative of ϕ with respect to x, $\frac{\partial\phi}{\partial x}$, (2.17) twice, along with the relation between η, γ, and η (2.20), and we have substituted the integration over x by the integration over ϕ in (2.25). To summarize this derivation, we insert the relation between the model parameters in Kobayashi's notation ϵ, γ, and τ and the physical parameters M^ϕ, σ, and η into (2.15) to arrive at the final phase-field equation with a double-well potential:

$$\frac{\partial\phi}{\partial t} = M^\phi \left\{\sigma\left[\nabla^2\phi - \frac{72}{\eta^2}\phi(1-\phi)\left(\frac{1}{2}-\phi\right)\right] - \frac{6}{\eta}\phi(1-\phi)\Delta g_0\right\}. \tag{2.28}$$

Going back to our original problem, to solve the propagation of a wave front with constant speed v_0 (2.2), we now can prove (2.3) just by inserting the second derivative (2.18). In fact, both terms cancel if the front is in the right contour. The first derivative (2.17) then motivates the ansatz for the driving force $m(\phi) \propto \frac{\partial}{\partial x}\phi \propto \phi(1-\phi)$ (2.9). Now we see that the hyperbolic transport Eq. (2.2) is turned into a parabolic equation with the second derivative in the space coordinate x. This type of equation—also termed a diffusion equation—is well behaved and easily solved on a computer. You can try this out using the program with which you tried to solve the hyperbolic Eq. (2.2).

2.4   Interpretation of the Phase-Field Equation

Let us now have a closer look at the phase-field Eq. (2.28) to determine the effect
of each term on the phase-field profile. The first term is the Laplacian, or diffusion
operator. This smoothens any profile, ensuring a smooth phase-field profile between
0 and 1. For the right-hand hyperbolic tangent profile, it contributes negative
increments above $\phi > \frac{1}{2}$ and positive increments below $\phi < \frac{1}{2}$, as indicated in
Fig. 2.1a. The second term stems from the double-well potential. Its first derivative
changes its sign at $\phi = \frac{1}{2}$ because the potential function is symmetric around $\frac{1}{2}$. Now
we see that the potential contribution has negative increments where the Laplacian
has positive increments, and vice versa [Fig. 2.1b].

This competition between the two contributions guarantees the stable phase-
field profile! If a numerical solution deviates from the correct profile, the
contributions will not balance but push the contour back to the correct contour.
Later, in Chap. 3, we will see that this balance is also violated for a curved interface
in more than one dimension. This will be used to consider capillarity. Furthermore,
Δg will not be a constant in real applications: it will be positive or negative, and
it will vary from point to point in space and time. It will depend on transport of
temperature, solute, and momentum, as will be explained in later chapters.

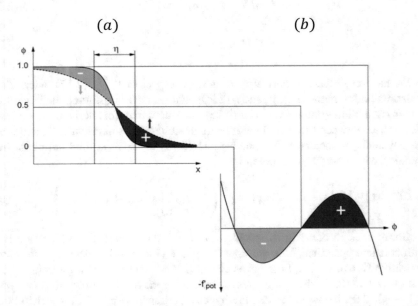

Fig. 2.1 The competition between spreading and contraction of the phase-field contour. Description see text

2.5   Phase-Field Equation and Traveling-Wave Solution for the Double-Obstacle Potential

As a general comment, we shall keep in mind that the potential function (2.8), which is called the double-well potential, is not a unique choice for any system. It was first introduced by Landau to explain the behavior of a ferromagnet at the critical point, which is called the Curie point in this system. Only at this critical point, which is a second-order phase transformation, is it justified to truncate the potential describing the interface energies—the so-called Landau potential—to this simple form (for details, see [6]). In phase field, however, we are dealing with a first-order phase transformation such as solidification or solid-state phase transformations, which are far away from a critical point. We also stick to the notion of a mesoscopic phase-field method (Sect. 1.4) whereby phenomena inside the interface cannot, and need not, be interpreted physically. Therefore, we have some freedom to choose a potential function in a different form than (2.8).

For reasons that will be explained in Chap. 6, the multi-phase-field method, as implemented in **OpenPhase** [3], applies the so-called double-obstacle potential:

$$f^{DO} = \frac{\gamma^{DO}}{2}\, |\phi(1-\phi)|\,. \tag{2.29}$$

The "obstacle" is realized by the absolute signs, which flip the negative branches of the function $\phi(1-\phi)$ to positive values.[4] Figure 2.2 shows a comparison between the double-well and double-obstacle potentials.

Fig. 2.2 The "double-well" and "double-obstacle" potentials

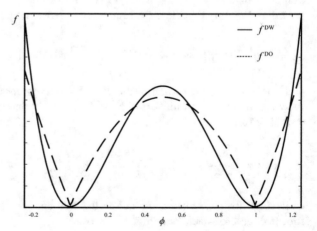

---

[4] Note that these absolute signs are omitted in several publications for better readability, but they are definitely necessary for the model to work. In the numerical calculations, they are realized by a sharp cutoff against negative values of $\phi < 0$ as well as against values larger than 1, $\phi > 1$.

As we can see, the graphs of the two potentials are similar. They are determined such that: (i) the free energy goes to infinity $f \to \pm\infty$ for $\phi \to \pm\infty$; (ii) they have two minima at $f = 0$ for $\phi = 0$ and $\phi = 1$; and (iii) the region of the so-called potential barrier between 0 and 1 has an area that is $\frac{1}{2}$ the interface energy σ. Other potentials can be used that fulfill above criteria, but they are little discussed in the phase-field literature. As long as we are speaking about a two-phase system, they should not show another minimum. More complex potentials, so-called "multi-well potentials" are used for multi-phase systems (e.g., [1, 4]).

The minimum solution of the free energy with the double-obstacle potential (2.29) is:

$$\phi^{DO}(x, t) = \begin{cases} 1 & \text{for } x - v_0 t < -\frac{\eta}{2}, \\ \frac{1}{2} - \frac{1}{2}\sin\left(\frac{\pi(x-v_0 t)}{\eta}\right) & \text{for } -\frac{\eta}{2} \leq x - v_0 t < +\frac{\eta}{2}, \\ 0 & \text{for } x - v_0 t \geq +\frac{\eta}{2}. \end{cases} \quad (2.30)$$

As before, we derive in the transition region $0 < \phi < 1$ (do this as an exercise!):

$$\frac{\partial \phi}{\partial x} = -\frac{\pi}{\eta}\sqrt{\phi(1 - \phi)}, \quad (2.31)$$

$$\frac{(\partial \phi)^2}{\partial x^2} = \frac{\pi^2}{\eta^2}\left(\frac{1}{2} - \phi\right). \quad (2.32)$$

The final task is to find a proper coupling function m^{DO} corresponding to the first derivative in ϕ, (2.31), of the phase-field profile (2.30) for the double-obstacle potential. We only give here the function and leave the proof as an exercise:

$$m^{DO} = \frac{1}{4}\left[(2\phi - 1)\sqrt{\phi(1 - \phi)} + \frac{1}{2}\arcsin 2\phi - 1)\right]\Delta g_0. \quad (2.33)$$

Repeating the analysis for the double-well potential, we find the relations between the model parameters ϵ^{DO}, γ^{DO}, and τ^{DO} and the physical parameters η, σ, and M^ϕ:

$$\eta = \pi\sqrt{\frac{\epsilon^{DO}}{\gamma^{DO}}}; \quad \sigma = \frac{\pi^2}{8}\frac{\epsilon^{DO}}{\eta} = \gamma^{DO}\frac{\eta}{8}; \quad M^\phi = \frac{\pi^2}{8}\frac{\eta}{\tau^{DO}}. \quad (2.34)$$

Collecting all these pieces, we define the free-energy density in physical notation for the double-obstacle potential:

$$f = \frac{8}{\pi^2}\frac{\sigma}{\eta}\left[\eta^2(\nabla\phi)^2 + \pi^2|\phi(1 - \phi)|\right] + \Delta g. \quad (2.35)$$

The normalization of the interface contribution $\frac{8}{\pi^2}$ is necessary to fulfill the condition that the integral of this term in the normal direction through the interface becomes the interface energy σ [cf. (2.26)]. We also chose the notation where σ is divided by a length, the interface width η, to emphasize that this contribution is an energy density like Δg. The term in the brackets is now a dimensionless contour function indicating the interface position. In the physical parametrization, the phase-field equation with the double-obstacle potential reads:

$$\frac{\partial \phi}{\partial t} = M^\phi \left\{ \sigma \left[\nabla^2 \phi - \frac{\pi^2}{\eta^2} \left(\frac{1}{2} - \phi \right) \right] - \frac{\pi}{\eta} \sqrt{\phi(1-\phi)} \Delta g_0 \right\}. \tag{2.36}$$

2.6   Gibbs–Thomson Limit of the Phase-Field Equation

Summarizing all contributions, we can return to the underlying physical equation of interest: the Gibbs–Thomson equation (1.5). We have already done this in 3D, relating the velocity of the interface $\vec{v}$ to the rate change of the phase field with the normal vector to the interface $\vec{n} = \frac{\vec{\nabla}\phi}{|\vec{\nabla}\phi|}$:

$$\dot{\phi} = \sum_{i=1}^{3} \frac{\partial \phi}{\partial x_i} \frac{\partial x_i}{\partial t} = \vec{\nabla}\phi \vec{v},$$

$$\vec{v} = \frac{\dot{\phi}}{|\vec{\nabla}\phi|} \vec{n}. \tag{2.37}$$

The curvature of the interface κ, which is given by the strange expressions in Fig. 2.3, will be derived in Chap. 3.

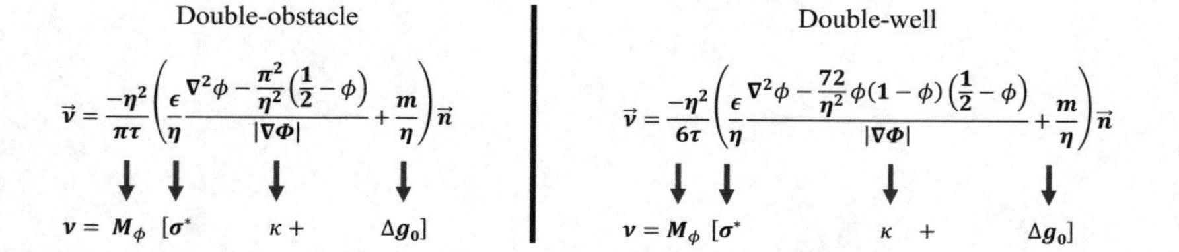

Fig. 2.3 The relations between the phase-field equations in Kobayashi's notation and the Gibbs–Thomson equation in physical notation for the double-well and double-obstacle potentials

2.7   Exercises

Exercise
Prove the first and second derivatives of the traveling-wave solutions for the double-well and double-obstacle potentials (2.17), (2.18) and (2.31), (2.32).

Exercise
Prove that the coupling function in (2.33) does the job, i.e., that the derivative of this function with respect to ϕ is proportional to the first derivative of ϕ in the case where a double-obstacle potential is used.

Exercise
Prove the relation (2.34) between the model parameters and the physical parameters for the double-obstacle potential.

Exercise
Check the relations between the model parameters as indicated in Fig. 2.3.

Further Reading
- Appendix A in the review "Phase-field models in materials science" [5], where another potential function, the "top hat" function, is introduced.

References

1. R. Folch, M. Plapp, Quantitative phase-field modeling of two-phase growth. Phys. Rev. E **72**, 011602 (2005). https://doi.org/10.1103/PhysRevE.72.011602
2. V.G. Kiselev, Y.M. Shnir, A.Y. Tregubovich, *Introduction to Quantum Field Theory* (CRC Press, Boca Raton, 2000). https://doi.org/10.1201/b16984
3. OpenPhase. https://openphase.rub.de/
4. O. Shchyglo, U. Salman, A. Finel, Martensitic phase transformations in Ni–Ti-based shape memory alloys: the Landau theory. Acta Mater. **60**(19), 6784–6792 (2012). https://doi.org/10.1016/j.actamat.2012.08.056
5. I. Steinbach, Phase-field models in materials science. Model. Simul. Mater. Sci. Eng. **17**, 073001 (2009)
6. I. Steinbach, Phase-field model for microstructure evolution at the mesoscopic scale. Ann. Rev. Mater. Res. **43**, 89–107 (2013). https://doi.org/10.1146/annurev-matsci-071312-121703

Chapter 3
Capillarity

3.1 Curvature of a Phase-Field Contour

In the previous chapter, the meanings of individual terms, the gradient and potential contributions of the free-energy functional [Eqs. (2.7) and (2.8), respectively], and their functional derivatives were discussed. In the free-energy functional, both terms contribute $\frac{1}{2}$ of the interface energy σ. In the phase-field equation of motion, either for the double-well potential (2.28) or double-obstacle potential (2.36), they cancel to 0 in 1D for the correct analytical solution. If the numerical solution deviates from the analytical solution, either because one has started from wrong initial conditions or because of numerical errors, the two terms act to push the interface to the right solution, i.e., they stabilize the interface contour. For a curved interface in 3D, it has already been mentioned that these terms do not cancel! They have the property to evaluate "curvature." This will be proven in this third chapter.

To summarize: the phase-field equation has two important properties: (i) to propagate the interface with a velocity related to the bulk free-energy difference Δg between different phases such that the energetically lower phase will grow and the other phase will shrink; (ii) to correct this transformation process for "capillarity" related to the local curvature of the interface. The latter also considers interface-energy anisotropy, which is a dominant effect in many metallurgical processes. This will be detailed in the second part of this chapter. Let us first prove the relation between the surface terms in the phase-field equation and curvature.

The formal mathematical definition of the curvature κ of any vector field is the divergence of the normal vector $\vec{n}$:

$$\kappa = \vec{\nabla}\vec{n} = \vec{\nabla}\frac{\vec{\nabla}\phi}{\left|\vec{\nabla}\phi\right|}. \tag{3.1}$$

© The Author(s) 2023
I. Steinbach, H. Salama, *Lectures on Phase Field*,
https://doi.org/10.1007/978-3-031-21171-3_3

Of course, we evaluate this for our phase field ϕ. Applying the divergence to the numerator and denominator of $\frac{\vec{\nabla}\phi}{|\vec{\nabla}\phi|}$ gives:

$$\kappa = \frac{1}{|\vec{\nabla}\phi|}\nabla^2\phi - \frac{\vec{\nabla}\phi}{|\vec{\nabla}\phi|^2}\vec{\nabla}\left|\vec{\nabla}\phi\right|$$

$$= \frac{1}{|\vec{\nabla}\phi|}\left[\nabla^2\phi - \frac{\vec{\nabla}\phi}{|\vec{\nabla}\phi|}\vec{\nabla}\left|\vec{\nabla}\phi\right|\right] \tag{3.2}$$

$$= \frac{1}{|\vec{\nabla}\phi|}\left[\nabla^2\phi - \vec{n}\,\vec{\nabla}\left|\vec{\nabla}\phi\right|\right]$$

$$= \frac{1}{|\vec{\nabla}\phi|}\left[\nabla^2\phi - \frac{\partial^2}{\partial n^2}\phi\right] \tag{3.3}$$

$$= \frac{1}{|\vec{\nabla}\phi|}\left[\nabla^2\phi - \frac{\pi^2}{\eta^2}(\frac{1}{2} - \phi)\right]. \tag{3.4}$$

We have inserted the definition of $\vec{n}$ into Eq. (3.2). This projects the divergence $\vec{\nabla}\left|\vec{\nabla}\phi\right|$ onto the normal through the interface, so this operator can be replaced by the second derivative in the normal direction in Eq. (3.3). We know this second derivative from the 1D analysis, Eqs. (2.18) or (2.32) for the double-well or double-obstacle potentials, respectively. In the last Eq. (3.4), the relation for the double-obstacle potential is inserted. The relation for the double-well potential works analogously. We see that this is exactly the expression used in the previous chapter for the evaluation of the capillarity term in the Gibbs–Thomson equation (cf. Fig. 2.3).

3.2 Anisotropy of Interface Energy

In the previous chapter, dealing with the analytic relation of a phase-field equation and the phenomenological Gibbs–Thomson equation for interface motion including capillarity, all entities—i.e., the phase-field mobility M^ϕ, the interface energy σ, the interface width η, and the thermodynamic driving force Δg—were taken as constants. This is, in general, not the case; sometimes it is a strong oversimplification. The mobility and the interface energy are anisotropic functions of the orientation of the interface with respect to the orientation of the adjacent phases. One distinguishes so-called "inclinations" of the interface with respect to the crystallographic orientation of individual crystal phases, and the misorientation of these phases if they are crystalline solids. In general, we have to specify three misorientation angles and two inclinations: five angles in total. This poses a challenge for determining these functions, but it also generates challenges for phase-field models in terms of coping with these functions.

The mobility M^ϕ is relatively easy; it is simply a scalar proportionality factor between time t and the so-called kinetic driving force acting on the interface, i.e., the "capillarity-corrected bulk free-energy difference." It will, in general, vary from point to point in the interface, because the misorientation and inclination angles of a curved interface vary along the interface. Depending on the physical model used for the mobility, it may cause numerical problems regarding time-stepping schemes (because it relates to time). However, from a conceptual point of view, it is easy and does not need further discussion here. In contrast, the interface energy is more difficult.

3.2.1 Interface-Energy Anisotropy: Phenomenological Picture

Interface-energy anisotropy is a conceptual challenge. For phenomenology, we will only discuss the simple case of a crystalline inclusion in an amorphous matrix, say a liquid. Then, the dependence of the interface energy on misorientation (three angles) vanishes. We will only consider two dimensions; then, only one inclination angle θ with respect to the crystal axis θ_0 remains (see Fig. 3.1). The Gibbs–Thomson equation [see Chap. 1, Eq. (1.5)] becomes:

$$\vec{v} = \vec{n} M \left(\sigma^* \kappa + \Delta g \right). \tag{3.5}$$

Here, $\sigma^* = \sigma + \sigma''$, with $\sigma'' = \frac{\partial^2}{\partial \theta^2} \sigma$, is called the "interface stiffness," i.e., the resistivity of the interface to bending. σ'' is also called the "Herring torque," as it was introduced by Herring in 1952 [3]. This tells us that the interfacial energy of a precipitate can be reduced by two separate mechanisms: (i) that the interface area is reduced; and (ii) that the interface is turned into a low-energy inclination. In fact, for the equilibrium shape of the precipitation surrounded by its melt, the interplay of both mechanisms has significant consequences. Equilibrium here is defined as the chemical potential over the system being constant. Furthermore, the precipitate neither grows nor shrinks, i.e., $\vec{v} = 0$. Since the precipitate has a curved interface, there is a potential jump of $\sigma^* \kappa$ between the bulk energy of the precipitate and the matrix, the so-called "capillarity pressure."

The Gibbs–Thomson equation (1.5) becomes ($\vec{v} = 0$):

$$0 = \sigma^* \kappa + \Delta g$$

$$\sigma^* \kappa = -\Delta g = \text{constant}. \tag{3.6}$$

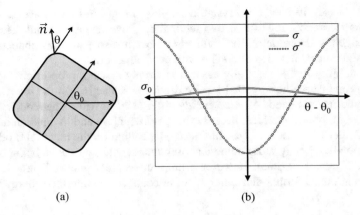

(a)   (b)

Fig. 3.1 (**a**) Sketch of the equilibrium shape of a crystal with strong interface anisotropy. The preferred crystallographic axis is rotated by θ_0 from the x axis. The surface normal at one specific point at the rounded corner is tilted by θ from the preferred axis. (**b**) The interface energy σ and interface stiffness σ^* as functions of $\theta - \theta_0$

If $\sigma^* \kappa = $ constant all along the interface, then σ^* has to be inversely proportional to κ, or, because σ^* is a material constant, κ has to be inversely proportional to σ^*. Where the stiffness is high, the curvature of the interface has to be low. If the stiffness is low, the interface can be easily bent, and we have edges or corners. We also know that the interface energy is high at edges and corners, so the interface area must be small. As a take-away message: where the interface energy is high, the stiffness is low, and vice versa. All of this is "general physics", not specific to phase field. It is the classical picture as drawn by Wulff [10], Herring [3], and others. But phase field incorporates this physics naturally. In the phase-field literature about dendritic growth, a simple anisotropy function with fourfold symmetry, in which the strength of the anisotropy is δ, is popular:

$$\sigma(\theta) = \sigma_0(1 + \delta \cos[4(\theta - \theta_0)]) \tag{3.7}$$

$$\sigma^*(\theta) = \sigma_0(1 + \delta \cos 4(\theta - \theta_0) - 16\delta \cos 4(\theta - \theta_0))$$

$$= \sigma_0(1 - 15\delta \cos 4(\theta - \theta_0)). \tag{3.8}$$

Before ending this section, let us make several remarks:

Firstly, we see that for $\delta > \frac{1}{15}$, the stiffness becomes negative (for $\theta = 0$). Phenomenologically speaking, inclinations of the interface that have a negative stiffness are forbidden. They collapse into edges and corners of the crystal: the crystal surface has "missing angles." In phase fields, this is not possible to realize, since, due to the diffuseness of the interface, edges and corners are always rounded off with a maximum curvature of the order of the reciprocal interface width $\frac{1}{\eta}$. Therefore, one has to "regularize" the corresponding function of the interface stiffness with respect to orientation, cutting it to a positive minimum value of the stiffness. We regard this as "technical."

Secondly, in 3D, i.e., having two inclination angles, one usually expands the interface energy and the stiffness in spherical harmonics. For a strongly anisotropic interface energy, e.g., for faceted crystals, other models may be appropriate (see Example at the end of this chapter).

Thirdly, in a solid, we also need to consider the three misorientation angles for each inclination-angle landscape.

Finally, at the atomistic scale, i.e., for nanocrystals, all these phenomenological relations—and also their microscopic or mesoscopic phase-field equivalents—will break down! The description of "capillarity" by curvature is only a statistical description of the atomic movement at interfaces. Atoms do not care about the curvature that we measure at an interface at a scale large compared to the atomic scale. But collectively they behave as if they would follow these rules. And there are other effects involved in the atomic motion connected to interfaces such as coupled-motion [1] or stress–strain effects at the nanoscale [2, 8]. Figure 3.1a schematically explains the meaning of the inclination angle with respect to the crystallographic axis and the normal vector of the interface in 2D. Figure 3.1b displays a simple model for the interface energy and stiffness as functions of the inclination angle.

3.2.2 Interface-Energy Anisotropy: Phase-Field Picture

In phase-field theory, the whole issue of interface-energy anisotropy is considered naturally, though there currently remain many open issues of technical character: it is not so easy. Let us start from the beginning, the phase-field equation as the functional derivative of the free energy, Eq. (2.6):

$$\frac{\delta\phi}{\delta t} \propto -\frac{\delta F}{\delta \phi}. \tag{3.9}$$

Here, we will only investigate the capillarity contribution, i.e., we will neglect the bulk energy difference between phases $\Delta g = 0$. In the physical notation, with anisotropic interface energy $\sigma(\vec{n})$ as a function of the normal vector to the interface $\vec{n}$ but with isotropic interface width η, the variation of the free energy then becomes:

$$\delta F = \int_\Omega d^3x \left\{ \eta\sigma(\vec{n})\,\delta\left[(\nabla\phi)^2 + \frac{\pi^2}{\eta^2}|\phi(1-\phi)|\right] \right. \tag{3.10}$$

$$\left. + \eta\,\delta\sigma(\vec{n})\left[(\nabla\phi)^2 + \frac{\pi^2}{\eta^2}|\phi(1-\phi)|\right]\right\}. \tag{3.11}$$

We use the double-obstacle potential; the case with the double-well potential works identically. In the first part (3.10), the variation is applied to the gradient and potential contributions. This part we already know from Chap. 2, Eq. (2.15). The new part is the variational derivative of the interface energy (3.11) since it is anisotropic with respect to inclination, and the interface normal $\vec{n}$ is a function of the gradient of the phase field $\vec{n} = \frac{\vec{\nabla}\phi}{|\vec{\nabla}\phi|}$. First, we notice that the term in the square brackets of Eq. (3.11) is a positive function of ϕ and can be approximated, neglecting higher-order curvature terms and using relation (2.31):

$$\left[(\nabla\phi)^2 + \frac{\pi^2}{\eta^2} |\phi(1-\phi)| \right] \approx 2(\nabla\phi)^2. \tag{3.12}$$

A general discussion of the variation $\delta\sigma(\vec{n})$ in 3D is quite involved, so we restrict ourselves here to 2D for the sake of tractability. Setting the rotation angle to zero $\theta_0 = 0$, we define the angle of the surface normal $\theta = \arctan(\frac{\phi_y}{\phi_x})$ with the derivative of ϕ in the x and y directions of a Cartesian coordinate system. Then we expand:

$$\delta\sigma(\theta)\,(\nabla\phi)^2 = \frac{\partial\sigma(\theta)}{\partial\theta}\delta\theta(\nabla\phi)^2 \tag{3.13}$$

$$= \frac{\partial\sigma(\theta)}{\partial\theta}\frac{\phi_x\delta\phi_y - \phi_y\delta\phi_x}{(\nabla\phi)^2}(\nabla\phi)^2 \tag{3.14}$$

$$= \frac{\partial\sigma(\theta)}{\partial\theta}\left(\phi_x\delta\phi_y - \phi_y\delta\phi_x\right) \tag{3.15}$$

$$= -\left[\frac{\partial}{\partial y}\left(\frac{\partial\sigma(\theta)}{\partial\theta}\phi_x\right)\delta\phi - \frac{\partial}{\partial x}\left(\frac{\partial\sigma(\theta)}{\partial\theta}\phi_y\right)\delta\phi \right]$$

$$= \frac{\partial^2\sigma(\theta)}{\partial\theta^2}\left(\frac{\partial\theta}{\partial y}\phi_x - \frac{\partial\theta}{\partial x}\phi_y\right)\delta\phi$$

$$\approx \frac{\partial^2\sigma(\theta)}{\partial\theta^2}|\nabla\phi|\,\kappa\,\delta\phi. \tag{3.16}$$

We leave the proof of this relation as an exercise. To help with this: we have three different differentials to consider in expanding the variation $\delta\sigma(\theta)$. Be reminded that all these differentials are linear operators that commute. In Eq. (3.13), the chain rule is used to expand the variation of σ into a differential in θ and a variation in θ. Then, in Eq. (3.14), the variation in θ is expanded in differentials in ϕ_x and ϕ_y, which by themselves are differentials in space. Then, partial integration is applied to the last term in Eq. (3.15) to respectively integrate the variations $\delta\phi_x$ and $\delta\phi_y$ in space to separate the single variation $\delta\phi$ (again neglecting boundary integral terms). This needs a lot of bookkeeping, but it is more or less straightforward. The final step $\frac{\partial\theta}{\partial y}\phi_x - \frac{\partial\theta}{\partial x}\phi_y \approx |\nabla\phi|\,\kappa$ is an approximation to lowest order in κ, where the variation of the angle along one Cartesian direction is weighted by the phase-field gradient in the other direction. This variation, clearly, has to vanish for a planar interface where the curvature is 0.

Collecting all contributions, we end up with the desired expression for the evolution of the phase field (valid for $0 < \phi < 1$ and $\Delta g \neq 0$), the equivalent of (2.36):

$$\frac{\partial \phi}{\partial t} = M^\phi \left\{ (\sigma + \sigma'') \left[\nabla^2 \phi - \frac{\pi^2}{\eta^2} \left(\frac{1}{2} - \phi \right) \right] - \frac{\pi}{\eta} \sqrt{\phi(1-\phi)} \Delta g \right\}. \qquad (3.17)$$

Please note that throughout Chap. 2, the interface energy σ was taken as isotropic, so no torque σ'' was considered. In general, this should always be there, as interfaces in crystalline phases are always dependent on orientation and inclination, even if only weakly. Several aspects shall be remarked upon. The form of Eq. (3.17) may not be the optimal choice for a numerical solution, since the prefactor $\sigma^* = \sigma + \sigma''$ in front of the Laplacian can be strongly varying for strong anisotropies. If an explicit time-stepping scheme is used for the phase-field equation, this may lead to a significant reduction of the possible time steps and/or it may introduce instabilities in the front. An alternative is to evaluate the curvature of the interface locally and average it within a certain region. Then, one can use the Herring torque term σ'' as a smoothly varying driving force for the interface. Alternatively, one may directly discretize the variation $\delta \sigma(\vec{n})$, Eq. (3.11), as done, e.g., in [4]. A comparison of different approaches for dendritic growth is given in [5].

3.3 Exercises

Exercise
Repeat for yourself the relation of the mathematical curvature (3.1) with the corresponding phase-field expression (3.4). Consider that this relation is only true if the phase field is in the "right contour."

Exercise
Prove the derivation of the Herring torque from the phase-field functional (3.16).

Example—Equilibrium Shapes
Anisotropy of interface energy is important in materials because it significantly impacts the morphology of the grain surface, the kinetics of interface
migration, and the overall structure of the composite. According to the
conventional definition of interface-energy anisotropy, it can be described
as a function of the crystallographic orientation of the interface, which
results in the equilibrium forms of individual crystals given by the Wulff
construction [10]. We use an anisotropy model here, with the interface energy
being a function of the inclination angle θ only. The interface stiffness with
respect to the inclination angle is modeled as [7]:

$$\sigma_\alpha^* = \sigma_\alpha(\theta_\alpha) + \sigma''(\theta_\alpha) = \frac{\sigma^0 a^2}{\left(\sin^2(\theta_\alpha) + a^2 \cos^2(\theta_\alpha)\right)^{\frac{3}{2}}}. \tag{3.18}$$

Validation of the Equilibrium Shapes
The **proposed model** was employed in a 3D simulation box with 64^3 grid
points of an initially spherical grain inserted into the melt. The inclination
angle is computed using the interface normal of the phase field and different
basic sets of facet normals, as shown in Fig. 3.2.

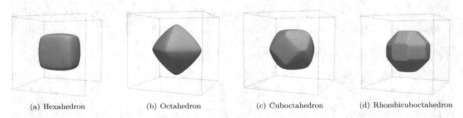

(a) Hexahedron   (b) Octahedron   (c) Cuboctahedron   (d) Rhombicuboctahedron

Fig. 3.2 Equilibrium shapes with different facet vectors: (**a**) only {100} facets; (**b**) only {111}
facets; (**c**) {100} and {111} facets; (**d**) {100}, {110}, and {111} facets

References

1. J.W. Cahn, Y. Mishin, A. Suzuki, Coupling grain boundary motion to shear deformation. Acta Mater. **54**(19), 4953–4975 (2006). ISSN: 1359-6454. https://doi.org/10.1016/j.actamat.2006. 08.004. https://www.sciencedirect.com/science/article/pii/S1359645406005313.
2. R. Darvishi Kamachali et al., Multiscale simulations on the grain growth process in nanostructured materials. Int. J. Mater. Res. **101**, 1332–1338 (2010). https://doi.org/10.3139/146.110419
3. C. Herring, The use of classical macroscopic concepts in surface-energy problems, in *Structure and Properties of Solid Surfaces*, ed. by R. Gomer, C.S. Smith. (University of Chicago Press, Chicago, 1952)
4. H.K. Kim et al., Phase-field modeling for 3D grain growth based on a grain boundary energy database. Model. Simul. Mater. Sci. Eng. **22**(3), 034004 (2014). https://doi.org/10.1088/0965-0393/22/3/034004
5. J. Kundin, I. Steinbach, Comparative study of different anisotropy and potential formulations of phase-field models for dendritic solidification. Comput. Mater. Sci. **170**, 109197 (2019). https://doi.org/10.1016/j.commatsci.2019.109197
6. G.B. McFadden et al., Phase-field models for anisotropic interfaces. Phys. Rev. E **48**, 2016–2024 (1993). https://doi.org/10.1103/PhysRevE.48.2016
7. H. Salama et al., Role of inclination dependence of grain boundary energy on the microstructure evolution during grain growth. Acta Mater. **188**, 641–651 (2020). https://doi.org/10.1016/j.actamat.2020.02.043
8. I. Steinbach, X. Song, A. Hartmaier, Phase-field model with plastic flow for graing rowth in nanocrystalline material. Philos. Mag. **90** (2010). https://doi.org/10.1080/14786430903074763
9. A.A. Wheeler, Cahn–Hoffman ξ-vector and its relation to diffuse interface models of phase transitions. J. Stat. Phys. **95**, 1245–1280 (1999)
10. G. Wulff, On the question of speed of growth and dissolution of crystal surfaces. Z. Krist. **34**(5/6), 449 (1901)

Chapter 4
Temperature

4.1 Thermodynamically Consistent Derivation of the Phase-Field Equation Coupled to Temperature

In the previous chapters, we studied the phase-field equation itself; it can be used to propagate planar waves, but also to study the effect of capillarity on the equilibrium structures of crystals. For phase transformations, however, we have to consider the release/consumption of latent heat and solute redistribution in an alloy. In this chapter, we will study the interplay between the growth of a crystal and the temperature field around it. We will do this for a solidification problem, but the situation is not restricted to solidification. During a solid-state transformation, the release of latent heat also has a significant effect. We will do this for a pure substance in which solute redistribution plays no role. However, in general, both fields—temperature and solute (see Chap. 5)—have to be coupled to the evolution of the phases in an alloy. See also "Example—Temperature Evolution During Rapid Solidification" at the end of this chapter.

The Gibbs free energy in the case that we solve the heat-conduction equation is not an appropriate starting point, since the Gibbs energy requires that the temperature be given. Instead, one starts from the entropy functional S. Following Wang et al. [5], the entropy functional S as the integral of the entropy density s over the domain Ω is defined by the internal energy density e and the free-energy density f:

$$S = \int_{\Omega} s = \int_{\Omega} \frac{e - f}{T}, \tag{4.1}$$

$$e = \rho c_p T + L(1 - \phi). \tag{4.2}$$

© The Author(s) 2023
I. Steinbach, H. Salama, *Lectures on Phase Field*,
https://doi.org/10.1007/978-3-031-21171-3_4

The internal energy of a solid is assumed to be linearly dependent on temperature T with constant density ρ, specific heat capacity c_p, and latent heat of fusion L. The governing equations for T and ϕ are derived consistently with the principle of entropy production $\dot{S} > 0$:

$$\tilde{\tau}\dot{\phi} = -\frac{\delta}{\delta\phi}\left(\int_\Omega \frac{f}{T}\right)_T = -\frac{1}{T}\left[\frac{\partial f}{\partial\phi} - \nabla\frac{\partial f}{\partial\nabla\phi}\right], \qquad (4.3)$$

$$\dot{e} = -\nabla M_T \nabla\frac{\delta}{\delta e}\left(\int_\Omega \frac{e}{T}\right)_\phi = -\nabla\left[M_T\nabla\frac{1}{T}\right]. \qquad (4.4)$$

Inserting the energy model (4.2) into (4.4), we obtain:

$$\dot{e} = \rho c_p \dot{T} - L\dot{\phi} = \nabla\frac{M_T}{T^2}\nabla T$$

$$\rho c_p \dot{T} = \nabla\lambda_T\nabla T + L\dot{\phi}, \qquad (4.5)$$

which is the well-known heat-conduction equation, with thermal conductivity $\lambda_T = \frac{M_T}{c_p T^2}$. The phase-field Eq. (4.3) is then almost identical to the previously derived equation, e.g., for the double-obstacle potential (2.36). The only modification is the prefactor $\frac{1}{T}$, which can be assimilated into the phase-field mobility M^ϕ. The thermodynamically consistent derivation of phase-field models is of special importance, because it enables the correlation of the model parameters with each other, as well as the establishment of a sound theoretical background in thermodynamics.

The derivation above defines a pair of coupled partial differential equations. Both of these are parabolic diffusion-type equations. The interesting feature is the source term in each equation: the release of latent heat in the heat-diffusion equation, which depends on $\dot{\phi}$, and the driving term for the phase field, which depends on the actual temperature of the interface T. This mutual coupling leads to the morphologically unstable evolution of a dendritic solid–liquid interface. It also poses a challenge for numerical solution schemes of this set of equations. Usually, they are solved explicitly in a staggered scheme, i.e., sequentially: one and then the other. In the literature, more advanced schemes can be found, but these do not prevail in general practice. A simple piece of code in C++, a typical 100-line program (not counting input and output) for a phase-field problem, is given in Appendix A.1 for you to try yourself.

4.2   Thin-Interface Limit

Let us now focus on one specific problem associated with a physically meaningful solution of the above-outlined solidification problem: solidification occurring at relatively high temperatures with high mobility of the solid–liquid interface M_{SL}^{ϕ}. Typically, this transformation is treated as "diffusion controlled," meaning that the interface T_i is set to capillarity-corrected thermal equilibrium at temperature $T_{\text{m}} + \frac{\sigma^* \kappa}{\Delta S_{\text{SL}}}$, in which $\kappa < 0$ for a convex dendrite tip and ΔS_{SL} is the entropy of fusion. This condition tells us that the Gibbs–Thomson equation, which we have used to control the speed of the interface, becomes undetermined. As a reminder, the Gibbs–Thomson equation reads:

$$v = M_{\text{SL}}^{\phi} \left(\sigma^* \kappa + \Delta S_{\text{SL}}(T_i - T_{\text{m}}) \right). \tag{4.6}$$

For any finite velocity v, the condition $M_{\text{SL}}^{\phi} \to \infty$ determines only the interface temperature $T_i = T_{\text{m}} - \frac{\sigma^* \kappa}{\Delta S_{\text{SL}}}$, but the velocity cannot be specified. To proceed in this situation, we consider the temperature profile ahead of a growing solidification front, as sketched in Fig. 4.1.

In the growing solid on the left-hand side of Fig. 4.1, the temperature can be assumed to be uniform. In the sharp-interface picture (blue line), there is a kink in the temperature profile, which relates to the release of heat at the solidification front. In this case, it helps that the kink of the temperature at the front can be related to the interface velocity (see the classical Stefan problem [4]):

$$Lv = \lambda_{\text{S}} \frac{\partial T}{\partial x}\bigg|_{\text{S}} + \lambda \frac{\partial T}{\partial x}\bigg|_{\text{L}} \approx \lambda \frac{\partial T}{\partial x}\bigg|_{\text{L}}, \tag{4.7}$$

in which $|_{\text{S}}$ and $|_{\text{L}}$ denote that the temperature gradient is evaluated in the solid or liquid, respectively. Fluxes in the solid can be neglected under steady-state conditions because the solid is uniformly at the melting temperature. Equation (4.7) is a balance equation between heat release due to solidification Lv and heat extraction due to diffusion into the supercooled liquid $\lambda_{\text{L}} \frac{\partial T}{\partial x}$. We can see that the velocity is now calculated from heat diffusion instead of being proportional to a thermodynamic driving force (Gibbs–Thomson condition with finite mobility). This is called the "diffusion-controlled limit."

In phase field, this limit seems difficult, because the phase-field equation for $\dot{\phi}$ corresponds to the Gibbs–Thomson equation. The seminal contribution of Karma and Rappel [1, 2] was to realize that the temperature-diffusion equation, or heat-conduction equation, can "easily" be integrated into the phase-field equation!

To understand this, we go back to Fig. 4.1: the dotted line indicates the phase-field contour, which is diffused over the interface width η; the green line indicates the temperature corresponding to this solution. The release of latent heat is no longer concentrated at the front (producing the kink in the temperature profile in the sharp-interface picture) but is continuous over the interface. Correspondingly, the temperature profile is smooth. However, outside the interface, we demand that both temperatures, the sharp-interface and diffuse-interface temperatures, match. This is called the "asymptotic matching condition," as applied to solving the temperature

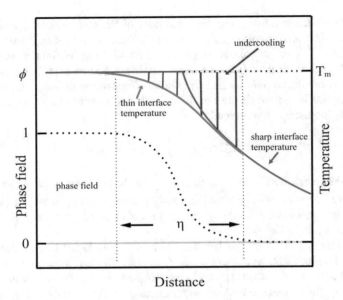

Fig. 4.1 Sketch of the temperature profile through an interface of a pure substance growing into an undercooled melt. The sharp front position is located at the center of the interface with a kink in the sharp-interface temperature. To the left is the solid at melting temperature T_m (neglecting capillarity effects for the moment); to the right is the undercooled melt, and the front is moving from left to right. The solid blue line depicts the temperature of the sharp-interface problem with a kink due to the release of latent heat at the moving front. The dotted line indicates the phase field, which is smeared out over the width η. The green line corresponds to the temperature profile of the phase-field model with a smeared-out release of latent heat within the interface. This shall match the sharp-interface solution outside the interface, but it has a systematic deviation within the interface. The latter produces a "spurious undercooling," which shall be used to compensate for "numerical undercooling" (see main text)

profile for a given phase-field solution analytically (in a 1D direction normal to the interface).

We can see directly that, systematically, the temperature of the phase-field model has to be lower (in the case of solidification into an undercooled melt) than the respective sharp-interface temperature. On the one hand, this is not good because it is a systematic error that cannot be avoided, and errors are never good; on the other hand, as long as the temperature outside the interface is correct, we will accept it. This is consistent with the notion that a mesoscopic phase-field model does not claim full physical resolution inside the interface. This systematic error then turns out to be very good from a practical point of view: it helps us to relate the velocity of the interface to undercooling!

There is no physical undercooling of the interface in the diffusion-controlled limit of a sharp-interface model, but there is necessarily a spurious undercooling in the case of a diffuse-interface model. If the right temperature solution as function of the velocity can be calculated analytically, it should coincide with the spurious undercooling of a good numerical solution $\Delta g^{\text{numerical}}$. For the analytical solution, please refer to the literature [1–3]. Here, we simply state that the spurious undercooling can be (on average over the interface) expanded to lowest order in the front velocity v as $\Delta g^{\text{spurious}} \approx Av$ with a constant A depending only on the materials parameter λ, ΔS_{SL}, and η. This can be understood easily because the spurious undercooling must vanish if the system reaches equilibrium, i.e., $v \to 0$. With little math, we can reformulate the Gibbs–Thomson equation:

$$v = M^\phi \left[\sigma^* \kappa + \Delta g \right] \tag{4.8}$$

$$= M^\phi \left[\sigma^* \kappa + \Delta g + 0 \right]$$

$$= M^\phi \left[\sigma^* \kappa + \Delta g + \Delta g^{\text{numerical}} - \Delta g^{\text{spurious}} \right].$$

Here, again, we add an "intelligent 0", $0 = \Delta g^{\text{numerical}} - \Delta g^{\text{spurious}}$. We assume that our numerics are good and that the numerical solution of the "thin interface temperature" within the interface matches the correct analytical solution (green line in Fig. 4.1). Then the numerical driving force can be read from the numerical solution and the spurious driving force can be handled analytically as a "spurious velocity". This we do!

$$v(1 + M^\phi A) = M^\phi \left[\sigma^* \kappa + \Delta g + \Delta g^{\text{numerical}} \right],$$

$$v = \frac{M^\phi}{1 + M^\phi A} \left[\sigma^* \kappa + \Delta g + \Delta g^{\text{numerical}} \right]. \tag{4.9}$$

This defines the effective mobility of a diffuse-interface model, $M_{\text{eff}}^\phi = \frac{M^\phi}{1+M^\phi A}$. It is largely determined from diffusion in the dying phase. This means that a phase-field model, coupled to temperature diffusion, runs with a spurious undercooling and a finite effective mobility, even if the physical mobility $M^\phi \to \infty$. It also matches the sharp-interface solution outside the "thin" interface. This is called "thin-interface asymptotic." The driving force Δg^{eff} comprises the physical and the spurious (but necessary) numerical contributions.

4.3 Exercises

Exercise
Repeat (4.8) to (4.9) for yourself!

Example—Temperature Evolution During Rapid Solidification
In this example, the system is subjected to the solidification conditions of an
additive manufacturing process, i.e., a process with a very high cooling rate
and very high thermal gradients. In the coupled phase-field model, the heat
conduction equation,

$$\rho C_p \dot{T} = \nabla \cdot (\lambda \nabla T) + Q_{local}, \qquad (4.10)$$

is solved implicitly. Here, Q_{local} handles all kinds of heat sources, including
both the release of latent heat and external heat sources. From the simulation
results shown in Fig. 4.2, we can see the growth of dendrites in cellular form
owing to the very high temperature gradient and cooling rate. The line scan
of the temperature plot along the z axis highlights that the dendrite tip is the
hottest region during solidification. This is due to the release of latent heat
during phase transformation.

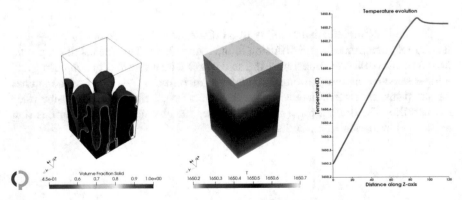

Fig. 4.2 Phase-field simulation results obtained for a solidification process under additive manufacturing conditions. Left: evolution of solid phase. Middle: temperature distribution at the box
surface in color coding for a given time step. Right: line scan of temperature along the z axis with
a peak temperature at the dendrite tip

Heat Release During Solidification

Further Reading

- The original publication on the thin-interface limit is by Alain Karma and Wouter-Jan Rappel [2]. The limit for a phase field coupled to solutal diffusion (see Chap. 5) is presented in [1]. Here, the so-called "anti-trapping current" for a model with a strong difference in the diffusion coefficients is also introduced. In alloy solidification, the liquid diffusivity is much larger than the solid diffusivity, and this has important consequences for treating the thin-interface limit.

References

1. A. Karma, Phase-field formulation for quantitative modeling of alloy solidification. Phys. Rev. Lett. **87**(11), 115701 (2001)
2. A. Karma, W.-J. Rappel, Quantitative phase-field modelling of dendritic growth in two and three dimensions. Phys. Rev. E **57**, 4323–4349 (1998)
3. I. Steinbach, Phase-field models in materials science. Model. Simul. Mater. Sci. Eng. **17**, 073001 (2009)
4. C. Vuik, Some historical notes on the Stefan problem. Nieuw Archief voor Wiskunde IV **11**(2), 157–167 (1993). ISSN: 0028-9825
5. S.-L.Wang et al., Thermodynamically-consistent phase-field models for solidification. Physica D **69**, 189–200 (1993)

Chapter 5
Concentration

The last chapter discussed the coupling of the phase field to temperature. You will remember that the phase-field functional has three contributions: the "gradient," the "potential," and the "bulk free-energy difference." The gradient has to fight with the other two contributions: the fight with the potential represents capillarity, and the fight of capillarity with the bulk free-energy difference determines phase transformations and growth or shrinkage of phases. Therefore, all these ingredients have to be considered simultaneously.

In this chapter, we consider the effect of solute on phase transformations. The treatment of solute within the diffuse interface of a phase-field model is one of the most controversial issues in phase-field theory. Here, we will give a condensed statement of the problem and then elaborate on the most recent model to couple phase-field theory and solute diffusion within the mesoscopic picture, the so-called "finite interface dissipation" (FID) model [12, 14].

5.1 Phase Field and Solute Concentration

The problem with treating the phase field and solute concentration in a consistent theory relates to the physical question of how crystal structure and solute composition correlate during a phase transformation. This is the chicken and egg problem: who comes first? During a phase transformation in an alloy, the crystal structure typically changes along with the solubility of the alloying elements. We can likely assume that both mechanisms go hand in hand, but a generally accepted rule cannot be given here. A special case is "spinodal decomposition," in which the crystal structure stays the same, but due to a miscibility gap, the solute partitions into two composition sets. The theory for the chemical interface energy that dominates in this case was proposed by Cahn and Hilliard [1]. This is now frequently applied in microscopic phase-field models of phase transformations

© The Author(s) 2023
I. Steinbach, H. Salama, *Lectures on Phase Field*,
https://doi.org/10.1007/978-3-031-21171-3_5

with structural components. This may work well if: (i) the structural part of the transformation is considered together with the chemical interface energy; and (ii) the interface is treated at the atomistic scale.

In a mesoscopic phase-field model, in which the interface width may reach the micrometer scale, a Cahn–Hilliard approach to solute diffusion will cause severe errors because the chemical interface energy in a Cahn–Hilliard model scales with the interface width [11]. If we consider the interface at the atomistic scale, with a roughly 1-nm width compared to a scaled interface width of $1\,\mu$m in a mesoscopic simulation, this error will be a factor of 1000. This situation is depicted in Fig. 5.1: the Gibbs energy of a two-phase system f^{chem} has a hump in the two-phase region between the equilibrium compositions c_α^{e} and c_β^{e} if the composition is treated as a continuous variable c between c_α^{e} and c_β^{e}. The integral over this hump in the normal direction through the interface gives the chemical interface energy $\sigma_{\text{chem}} = \int dn\ f^{\text{chem}} \propto \eta$.

There are three options to cope with this problem:

- Use adaptive methods to scale down the interface to the atomistic scale.
- Construct a Gibbs-energy landscape without a hump between the phase concentrations.
- Split the composition c into phase compositions c_α.

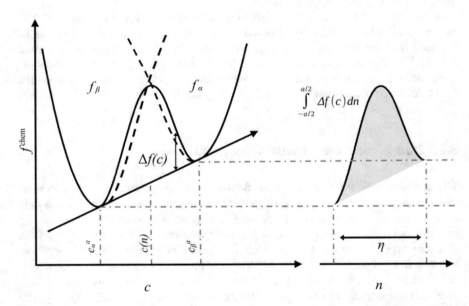

Fig. 5.1 Sketch of the Gibbs energy of a two-phase system (left). If we map the concentration between the phase concentrations c_α and c_β onto the normal n through the interface, the hump in the chemical Gibbs energy f^{chem} transfers to a region in space with increased energy (right). This region scales with the interface width η

The first approach, to use adaptive finite elements with mesh refinement towards the interface, raised some interest in the early days of the application of phase fields to solidification problems (see, e.g., [10]). This works sufficiently well in 2D; however, it is hardly possible to resolve structures in the micrometer range in 3D.

The second approach, as applied by Echebarria et al. to dendritic solidification of a binary model alloy [2], is difficult to generalize for an arbitrary Gibbs-energy landscape, and particularly for multicomponent materials. Nevertheless, it presents a rigorous procedure to get rid of spurious effects of coupling between solute redistribution and phase evolution in the interface.

The third approach is mostly used today for applications in multicomponent materials. First introduced by Tiaden et al. [13], this splits the overall composition c into the phase compositions c_α and c_β, where ϕ indexes the α phase:

$$c = \phi c_\alpha + (1 - \phi)c_\beta. \tag{5.1}$$

Hereby, the phase compositions c_α and c_β may deviate from the compositions in equilibrium c_α^e and c_β^e. The extra degree of freedom that arises from this construction is fixed by relating the phase compositions by a partition coefficient taken from the equilibrium phase diagram, as done in the original work [13]. A generalization with an extrapolation scheme to multicomponent phase diagrams is presented by Eiken et al. [3]. An alternative is to postulate local chemical equilibrium within the interface, as proposed by Kim et al. [5], i.e., equal chemical potential: $\mu_\alpha = \mu_\beta = \mu$. This is called the "KKS" model from the names of the authors, and it is quite popular. It has specific problems of a technical nature when applied to a general Gibbs-energy landscape. Therefore, Plapp developed the so-called grand-potential approach [8].

The grand potential is a Legendre transformation of the free energy, in which the chemical potential replaces the composition as a state variable. Now, if we assume equal chemical potential within the interface, there is only one variable, μ, *and* we are automatically in the minimum state of the free energy within the interface. Thereby, as a side product, the spurious interface energy is gone. Now one solves a diffusion equation for the chemical potential (instead of the diffusion equation for the composition) together with the phase-field equation. The diffusion equation for the chemical potential is formally identical to the diffusion equation for the temperature [Chap. 4, Eq. (4.5)] with a source term proportional to the rate change of the phase field ϕ. One difficulty remains: how to invert the chemical potential into the phase compositions, which are, of course, the variables of interest if we are investigating a phase transformation in an alloy. An oft-applied remedy is to approximate the Gibbs energies of the individual phases as parabolic. The chemical potential then becomes linear in concentration, and one can directly read the composition in the bulk phases from the chemical potentials.

One severe drawback of all the approaches described above is that they are restricted to local equilibrium within the interface, or at least to "close to local equilibrium." If this condition breaks down—e.g., in rapid solidification, where solute trapping appears, or in solid-state phase transformations at lower temperatures (such as bainitic or martensitic transformations)—one has to take a different approach.

5.2   Finite Interface Dissipation Model

The general idea of the FID model [12, 14] is to treat the phase compositions as separate variables within the interface that are connected by the conservation constraint of composition. Technically, one separates the diffusion fluxes into long-range fluxes within the bulk phases and into short-range redistribution fluxes between the individual phases within the interface. Then one can start from arbitrary initial phase-composition conditions; they shall converge toward local equilibrium within the interface if the kinetics of the process permits. We also assume that, in reality, the composition within the interface is not strictly in local equilibrium.

To build such a model, we start from the following assumptions:

- The phase compositions c_α within the interface are a constraint to the mixture composition $c = \sum_\alpha \phi_\alpha c_\alpha$. For a solid–liquid interface, this is simply $c = \phi_S c_S + \phi_L c_L$
- The phase concentrations are treated as independent variables, only subject to the sum constraint above.

In a mesoscopic model, the interface is taken as an effective reference volume without specifying its actual position and orientation. Within this reference volume, we take the two phases as "mixed" and allow them to exchange solute to lower the total Gibbs energy of the reference volume. We define the chemical free-energy density f^{chem} as a weighted sum of the chemical free energies of the individual phases $f_{\alpha/\beta}^{\text{chem}}$:

$$f^{\text{chem}} = \phi_\alpha f_\alpha^{\text{chem}} + \phi_\beta f_\beta^{\text{chem}} + \lambda \left(c - \phi_\alpha c_\alpha + \phi_\beta c_\beta \right), \qquad (5.2)$$

where the Lagrange multiplicator λ ensures conservation of the concentration *and* gives us the possibility to take the phase concentrations c_α fully as independent variables. Then, we define a redistribution flux between the phases from the demand of energy reduction with the permeability P (for details see [12]):

$$\phi_\alpha \dot{c}_\alpha = -P \frac{\delta}{\delta c_\alpha} F = -P \frac{\partial}{\partial c_\alpha} f^{\text{chem}} = -P \left[\phi_\alpha \frac{\partial f_\alpha}{\partial c_\alpha} - \phi_\alpha \lambda \right], \qquad (5.3)$$

$$\phi_\beta \dot{c}_\beta = -P \frac{\delta}{\delta c_\beta} F = -P \frac{\partial}{\partial c_\beta} f^{\text{chem}} = -P \left[\phi_\beta \frac{\partial f_\beta}{\partial c_\beta} - \phi_\beta \lambda \right]. \qquad (5.4)$$

The permeability is the inverse of the resistivity of the interface against redistribution between the different phases. We can certainly assume that the interface has an influence on the diffusion of solutes, that it hinders diffusion or even makes it easier because of free volume within the interface region. This will depend on the particular kind of interface. For a high permeability, it is easy to exchange solute between the phases. A low permeability will characterize a passivated interface through which no solute can pass.

From the conservation condition $\dot{c} = 0$, which shall hold within one reference volume without external fluxes (we will add these later), one determines the Lagrange multiplier:

$$
\begin{aligned}
0 = \left(\phi_\alpha c_\alpha + \phi_\beta c_\beta\right)^{\cdot} &= \dot{\phi}_\alpha c_\alpha + \dot{\phi}_\beta c_\beta + \phi_\alpha \dot{c}_\alpha + \phi_\beta \dot{c}_\beta \\
&= \dot{\phi}_\alpha c_\alpha + \dot{\phi}_\beta c_\beta + \phi_\alpha \dot{c}_\alpha \\
&\quad - P\left[\phi_\alpha \frac{\partial}{\partial c_\alpha} f^{\mathrm{chem}} + \phi_\beta \frac{\partial f_\beta}{\partial c_\beta} - (\phi_\alpha + \phi_\beta)\lambda\right].
\end{aligned}
\tag{5.5}
$$

$$
\lambda = \phi_\alpha \frac{\partial f_\alpha}{\partial c_\alpha} + \phi_\beta \frac{\partial f_\beta}{\partial c_\beta} - \frac{\dot{\phi}_\alpha c_\alpha + \dot{\phi}_\beta c_\beta}{P}.
\tag{5.6}
$$

Substituting λ into Eqs. (5.3) and (5.4), we get the evolution equations for each phase inside the reference volume,

$$
\phi_\alpha \dot{c}_\alpha = P\phi_\alpha \phi_\beta \left(\frac{\partial f_\beta}{\partial c_\beta} - \frac{\partial f_\alpha}{\partial c_\alpha}\right) + \phi_\alpha \dot{\phi}_\alpha (c_\beta - c_\alpha),
\tag{5.7}
$$

$$
\phi_\beta \dot{c}_\beta = P\phi_\alpha \phi_\beta \left(\frac{\partial f_\alpha}{\partial c_\alpha} - \frac{\partial f_\beta}{\partial c_\beta}\right) + \phi_\beta \dot{\phi}_\beta (c_\alpha - c_\beta).
\tag{5.8}
$$

Finally, we add the long-range diffusion in the bulk phases in the standard way. D_α and D_β are the diffusion coefficients, and $\mu_\alpha = \frac{\partial f_\alpha}{\partial c_\alpha}$ and $\mu_\beta = \frac{\partial f_\beta}{\partial c_\beta}$ are the chemical potentials:

$$
\phi_\alpha \dot{c}_\alpha = \vec{\nabla}(\phi_\alpha D_\alpha \vec{\nabla} c_\alpha) + P\phi_\alpha \phi_\beta (\mu_\beta - \mu_\alpha) + \phi_\alpha \dot{\phi}_\alpha (c_\beta - c_\alpha),
\tag{5.9}
$$

$$
\phi_\beta \dot{c}_\beta = \vec{\nabla}(\phi_\beta D_\beta \vec{\nabla} c_\beta) + P\phi_\alpha \phi_\beta (\mu_\alpha - \mu_\beta) + \phi_\beta \dot{\phi}_\beta (c_\alpha - c_\beta).
\tag{5.10}
$$

By construction, both equations overlap in the interface, where two mechanisms of solute redistribution are considered: redistribution due to a chemical-potential difference $\mu_\alpha - \mu_\beta \neq 0$ and renormalization if the fraction of phases changes $\dot{\phi}_\alpha = -\dot{\phi}_\beta \neq 0$.

Remark All phase-field implementations that use the splitting of the composition c into phase compositions c_α have to include a redistribution step and a renormalization step. This redistribution procedure is, unfortunately, never discussed in detail in the literature, but it is considered as a technical concern. Partitioning and redistribution is, however, a crucial consideration if you want to implement your own solution.

The diffusion Eqs. (5.9) and (5.10) for the phase compositions α and β converge to the KKS model in the case of a very high permeability $P \to \infty$ because the redistribution is then very fast and the chemical potentials μ_α and μ_β must become equal. In the case of a low permeability, however, the interface concentration and the partitioning in the interface may significantly deviate from local equilibrium. Currently, there is no general model to determine the permeability parameter for a particular interface. The model as presented in the original paper [12] must be considered to be in error, since it has an inverse dependence of the permeability on the interface width, which must be considered as nonphysical in a mesoscopic model. A special model for the permeability in an electrochemical system, derived from experiments, is given in [9].

The model has another intriguing feature regarding the influence of the permeability on the motion of the interface. Going back to the model for the chemical free energy (5.2), we notice first that the Lagrange parameter λ takes the role of an effective chemical potential of the interface. In fact, it is—according to (5.6)—a mixture of the chemical potentials of the individual phases, linear in ϕ. Furthermore, it is also a function of the rate change of the phase field $\dot{\phi}$! Since all kinetic-model equations—phase-field, temperature, and composition up to now (stress and strain will be added in Chap. 2)—are derived as thermodynamically consistent from the same free-energy functional F, this term will enter all kinetic equations. The phase-field equation [cf. Chap. 2, Eq. (2.36)] becomes:

$$\dot{\phi}_\alpha = -M^\phi \frac{\delta F}{\delta \phi}$$

$$= M^\phi \left\{ \sigma^* [\nabla^2 - \frac{\pi^2}{\eta^2}(\frac{1}{2} - \phi)] - \frac{\pi}{\eta}\sqrt{\phi(1-\phi)}\Delta g_{\alpha\beta}^\lambda \right\}, \qquad (5.11)$$

$$\Delta g_{\alpha\beta}^\lambda = f_\alpha - f_\beta - \lambda(c_\alpha - c_\beta)$$

$$= f_\alpha - f_\beta - \left(\phi\mu_\alpha + (1-\phi)\mu_\beta - \frac{\dot{\phi}(c_\alpha - c_\beta)}{P} \right)(c_\alpha - c_\beta). \quad (5.12)$$

Rearrangement of the terms proportional to $\dot{\phi}_\alpha$ finally leads to:

$$\dot{\phi}_\alpha = K \left\{ \sigma_{\alpha\beta}[\nabla^2\phi_\alpha + \frac{\pi^2}{\eta^2}(\phi_\alpha - \frac{1}{2})] - \frac{\pi^2}{8\eta}\Delta g_{\alpha\beta} \right\}, \qquad (5.13)$$

$$K = \frac{P\eta M^\phi}{P\eta + M^\phi(c_\alpha - c_\beta)^2}, \qquad (5.14)$$

$$\Delta g_{\alpha\beta} = f_\alpha - f_\beta + (\phi_\alpha\mu_\alpha + \phi_\beta\mu_\beta)(c_\beta - c_\alpha). \qquad (5.15)$$

The phase-field mobility is modified by a mechanism that is related to solute redistribution and diffusion. The interface mobility M^ϕ in the thin-interface limit should be taken as the effective mobility derived in Chap. 4, at least for high permeability P when the model converges to the KKS model. To date, a consistent treatment for the non-equilibrium case (small P) for thin interfaces is still missing.

We see that there are three distinct limits for the new phase-field mobility K:

- $P \to \infty$: $K \to M^\phi$;
- $P \to 0 \cup (c_\beta - c_\alpha) \neq 0$: $K \to 0$;
- $P \to 0 \cup (c_\beta - c_\alpha) = 0$: $K \to M^\phi$.

The first case, with high permeability, leaves the phase-field equation unchanged, and the ID model converges to the KKS model with equal chemical potentials within the interface. The second case represents a passivated interface: no redistribution of solute is allowed, but there is a jump in concentration between the phases. Any transformation is forbidden, and the transformation stops. The most interesting case is the third case: there is no concentration difference between both phases, as in a partitionless transformation like martensitic transformation, and the phase-field mobility is unaffected, i.e., the transformation can proceed with very high speed. Again, no detailed studies of this have been conducted to date.

5.3 Multicomponent Alloy Transformation

Up to here we were mainly speaking about a binary alloy with one composition variable c. This is important and gives you all the relevant information about "alloy transformation," including the general problems associated with it as partitioning and getting the phase diagram right. The extension to multicomponent alloys is straightforward (see [14]) from a theoretical point of view. It is more than "involved" from a practical point of view. Technical alloys, steels, nickel based superalloys, aluminum or copper alloys, brass and others, but also minerals and ceramics are multicomponent with 10 or more individual components. Also vacancies have to be considered for completeness. The first challenge which arises is the phase diagram from which we would like to read the equilibrium compositions of two particular phases, see Fig. 1.1: There is no way to display a phase diagram of a 10 component alloy and we need a numerical procedure to do the calculation of equilibria and deviation from equilibrium automatically.

The second challenge is related to redistribution and diffusion during a phase transformation. The individual components i of an n_c-component alloy in phase state α, are not independent but connected by constraint to the sum constraint:

$$\sum_{i=1}^{n_c} c_\alpha^i = 1 \qquad (5.16)$$

The third challenge, and a challenge in its own right, is the construction of the Gibbs energy landscape of a multicomponent alloy with 10s of possible phases. For the latter we have to rely on the so-called CALPHAD (CALculation of PHAse Diagrams) technique (see e.g. Lukas et al. [7]). Open or commercial databases are available which provide the Gibbs energy of a special phase (stable, metastable or even unstable) as function of temperature, pressure and all available components in a special machine readable format. These you simply insert into the phase-field functional (1.4) to calculate the Gibbs energy difference $\Delta g(\phi, T, c, \epsilon, \ldots)$. In the same way one can access the chemical potentials μ_α, μ_β in the Sect. 5.2. Now they are also multi-component $\mu_\alpha \rightarrow \mu_\alpha^i$ and one has to take the derivative in the directions i [14]. As said, in practice this may become very involved.

The last challenge to be mentioned here is multicomponent diffusion, again a challenge in its own right. We would like to incorporate atomic mobilities from so-called "mobility databases" which accompany thermodynamic databases. We reference here to "Further reading."

In the literature several approaches to cope with "multicomponent transformations" are established:

- Hard-coding of thermodynamic functions. This will be helpful if you have a low ranked problem, say $n < 4$, and a suitable analytic description of the Gibbs energy landscape available. Also you will need your own code and sufficient programming skills. And you will have to do the job for each new problem.
- Local parabolic approximation of the Gibbs energy landscape. Local means here in composition space for your given alloy system, temperature and pressure. The approximation you may do with the help of a CALPHAD software, or from a given analytic description. The approach has also the advantage that for this quadratic approximation the chemical potential is linear with the composition!
- Local linearization of the phase diagram [3]. This approach is somehow similar to the previous approach, but quite different in technical aspects. We consider it the most efficient approach for large scale computation, but it may not be suited for larger deviation from local equilibrium at the interface.
- Direct coupling to databases. This is the most general way of incorporating the full CALPHAD scheme into a phase-field calculation. Of course it has the highest computational cost. An example using the OpenPhase software is shown at the end of this chapter (Fig. 5.2).

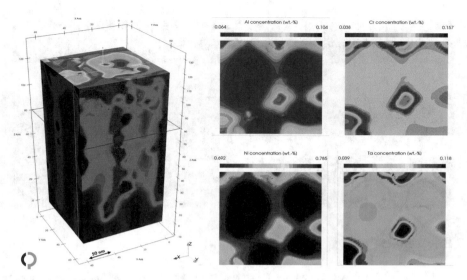

Fig. 5.2 Phase-field simulation of directional solidification of a four component Ni-Al-Cr-Ta superalloy under additive manufacturing conditions. Left: Al composition on the surface of the simulation box. Right: Composition profiles for the individual alloy component on a cross section horizontally through the simulation box (position indicated in left figure). Colour bars above each profile display the composition range. The directed dendrites under these conditions show almost no side branches and interdendritic γ' is precipitated at the end of solidification

5.4 Exercises

Exercise
Derive the expression for the Lagrange multiplier λ (5.6).

Example—Multicomponent Alloy Solidification
The following example relates to multicomponent alloy solidification in a Ni-based superalloy. A model alloy of the commercial alloy CMSX4 with four components, Ni, Al, Cr and Ta, is constructed such to yield a comparable fraction of secondary γ' phases at the solvus temperature of this alloy. The solidification conditions relate to additive manufacturing, as in the previous Example in Chap. 4.

Multicomponent Alloy Solidification

Further Reading

- Extrapolation scheme for multicomponent phase equilibria [3].
- Pair-wise multicomponent diffusion approach [4, 6]
- FID model for multicomponent alloy transformation [14] and sublattice ordering [15].

References

1. J.E. Cahn, J.E. Hilliard, Free energy of a nonuniform system. I. Interfacial free energy. J. Chem. Phys. **28**, 258–267 (1958)
2. B. Echebarria et al., Quantitative phase-field model of alloy solidification. Phys. Rev. E **70**, 061604 (2004)
3. J. Eiken, B. Böttger, I. Steinbach, Multiphase-field approach for multicomponent alloys with extrapolation scheme for numerical application. Phys. Rev. E **73**, 066122 (2006)
4. D. Gaertner, K. Abrahams, J. Kottke, V.A. Esin, I. Steinbach, G. Wilde, S.V. Divinski, Concentration-dependent atomic mobilities in FCC CoCrFeMnNi high-entropy alloys. Acta Mater. **166**, 357–370 (2019)
5. S.G. Kim, W.T. Kim, T. Suzuki, Phase-field model for binary alloys. Phys. Rev. E **60**(6), 7186–7197 (1999)
6. J. Kundin et al., Pairexchange diffusion model for multicomponent alloys revisited. Materialia **16** (2021). https://doi.org/10.1016/j.mtla.2021.101047
7. H. Lukas, S.G. Fries, B. Sundman, *Computational Thermodynamics: The Calphad Method* (Cambridge University Press, Cambridge, 2007)
8. M. Plapp, Unified derivation of phase-field models for alloy solidification from a grand-potential functional. Phys. Rev. E **84**, 031601 (2011) https://doi.org/10.1103/PhysRevE.84.031601
9. U. Preiss et al., A permeation model for the electrochemical interface. Model. Simul. Mater. Sci. Eng. **21**(7), 074006 (2013). https://doi.org/10.1088/0965-0393/21/7/074006
10. N. Provatas, N. Goldenfeld, J. Dantzig, Efficient computation of dendritic microstructures using adaptive mesh refinement. Phys. Rev. Lett. **80**, 3308–3311 (1998). https://doi.org/10.1103/PhysRevLett.80.3308
11. I. Steinbach, Phase-field model for microstructure evolution at the mesoscopic scale. Ann. Rev. Mater. Res. **43**, 89–107 (2013). https://doi.org/10.1146/annurev-matsci-071312-121703
12. I. Steinbach, L. Zhang, M. Plapp, Phase-field model with finite interface dissipation. Acta Mater. **60**, 2689–2701 (2012)
13. J. Tiaden et al., The multiphase-field model with an integrated concept for modeling solute diffusion. Physica D **115**, 73–86 (1998)
14. L. Zhang, I. Steinbach, Phase-field model with finite interface dissipation: extension to multicomponent multi-phase alloys. Acta Mater. **60**, 2702–2710 (2012)
15. L. Zhang et al., Incorporating the CALPHAD sublattice approach of ordering into the phase-field model with finite interface dissipation. Acta Mater. **88** (2015). https://doi.org/10.1016/j.actamat.2014.11.037

Chapter 6
Multi-Phase-Field Approach

6.1 Phase-Field Functional for Multiple Phases

Technical materials have multiple phases. In a CALPHAD (CALculation of PHAse Diagrams) database for iron alloys and steel, there are Gibbs-energy functions of several hundreds of different physical phases tabulated, and these are dependent on the crystal structure, composition, temperature, and pressure. According to Gibbs' phase rule, there can, however, only be $n = 2 + n_c$ stable phases in equilibrium, where n_c is the number of components of the alloy composition. For a steel, we may put $n_c = 10$, i.e., 12 different phases could be stable in equilibrium. Since, however, a steel is rarely in thermodynamic equilibrium but is more commonly in a kinetically stabilized off-equilibrium state with a locally varying composition, you may find 20 or more individual phases in a steel sample. In phase-field theory, we also attribute different grains—i.e., crystallites with the same crystallographic phase but a different orientation in space—to different phase fields. We do this to identify the interfaces between these grains, which will evolve in time (see Chap. 3). We can then have thousands of phase fields. How do we tackle this?

For completeness, there are two fundamentally different approaches for this "multi-grain" problem. The microscopic, or physical order-parameter models, use a free-energy landscape. Generally speaking, this is a special potential function with different minima that are dependent, for example, on the orientation of the crystallites [1, 11].

Another approach is the introduction of orientation as an order parameter [5, 6]. These approaches are generally considered as elegant from a theoretical point of view, but they are difficult to apply to practical problems. The reader should build his/her own opinion on this.

The third approach, which is more "brute force," is to address each crystallite, regardless of its phase, by its own phase field. We call this the "multi-phase-field approach" [12, 13]. Also popular is the so-called "vector order parameter" model by Fan and Chen [2]. Both models consider a set of N phase fields ϕ_α ($\alpha = 1 \ldots N$)

© The Author(s) 2023
I. Steinbach, H. Salama, *Lectures on Phase Field*,
https://doi.org/10.1007/978-3-031-21171-3_6

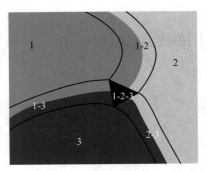

Fig. 6.1 Scheme of a triple junction between three phase fields. Within the bulk regions, only the information of the phase-field index has to be stored to identify the phase. Within the diffuse interface, two phase indices and one value of the phase field have to be stored. In junctions, we need the information of all phase-field indices, and all phase fields need to be stored. Of course, all of this information evolves dynamically during the calculation when the interfaces move

that can be attributed to different properties such as orientation, composition, crystal structure etc. We will describe here only the multi-phase-field approach [12, 13]. We distinguish three cases, as depicted in Fig. 6.1:

- Bulk: only one phase field has the value of 1, all other fields have the value of 0.
- Dual interface: two phase fields have values between 0 and 1, all other fields are 0.
- Junctions: we define a number $\tilde{N} \leq N$ of non-zero fields that overlap in a multiple junction. In Fig. 6.1 $\tilde{N} = 3$ in the triple junction in the center.

Within this lecture, we will only deal with "junctions," since the rest is known from dual phase fields as treated in the previous lectures. There is nonetheless enough to do...

The first thing to do is to define the free-energy density of a multi-phase-field model:

$$f = \sum_{\alpha=1}^{\tilde{N}} \left\{ \sum_{\beta=\alpha+1}^{\tilde{N}} \frac{8}{\pi^2} \frac{\sigma_{\alpha\beta}}{\eta} \left[-\eta^2 \nabla\phi_\alpha \nabla\phi_\beta + \pi^2 |\phi_\alpha\phi_\beta| \right] + \phi_\alpha f_\alpha^{\text{bulk}} \right\}. \qquad (6.1)$$

We sum over all $\tilde{N}$ phases/grains present within the junction for both the capillarity term, or interface, and the bulk free energy f_α^{bulk}, weighted by the local fraction of this phase α given by the phase field ϕ_α. This bulk free-energy density will be a function of temperature, concentration, stress, and strain, and it may contain magnetic or other contributions [cf. also Eq. (2.9) in Chap. 2].

The capillarity term is more complicated; it is expanded in pairs of phases, i.e., we have a second sum over all $\tilde{N}$ phases. Note that the second sum runs only from $\beta = \alpha + 1 \ldots \tilde{N}$ so as not to count contributions twice. There is also no diagonal term $\beta = \alpha$ because this form does not relate to an interface between two phases.

For triple junctions between three phase fields (usually a "triple line" in 3D), we find three pairs of phases. For quadruple junctions between four phases, we already have six pairs, and there are ten pairs for $\tilde{N} = 5$ and so on. The junctions are then modeled by the superposition of dual interface contributions with the interface energy of a pair $\sigma_{\alpha\beta}$. In general, we have $\tilde{N}$ combinatorial 2 interfaces for $\tilde{N}$ phases.

There may be higher-order contributions, i.e., contributions where more than two phase fields and their gradients are considered simultaneously. They should give a special penalty to multiple junctions, as discussed by Miyoshi and Takaki [7]. The present form (6.1) is just a pragmatic setting whereby all parameters are, in principle, well-defined physical entities. The interface width η is not considered a physical entity at the mesoscopic scale and is set to a constant value. The interface density $\left[-\eta^2 \nabla\phi_\alpha \nabla\phi_\beta + \pi^2 |\phi_\alpha\phi_\beta|\right]$ is expanded in pairs as a pragmatic choice. One can easily check (and please do so) that this form reduces in the case of a dual interface $\tilde{N} = 2$ to the standard form with a double-obstacle potential (2.35) where the bulk free energy is weighted linearly in ϕ_α.

6.2   Double Obstacle versus Double Well in Multi Phase Field

You may skip this part on first reading, but it is important! Why do we choose a double-obstacle potential instead of a double-well potential? An important feature of the multi-phase-field theory [12] is the sum constraint in junctions: $\sum_{\alpha=1}^{\tilde{N}} \phi_\alpha = 1$. For the double-well potential we write the potential (without prefactors) f_0^{DW}:

$$f_0^{DW} = \sum_{\alpha=1...\tilde{N}, \beta=\alpha+1...\tilde{N}} \phi_\alpha^2 \phi_\beta^2. \tag{6.2}$$

The maximum of the potential in the center of the junction, where $\phi^\alpha = \phi^\beta = \ldots = \dfrac{1}{\tilde{N}}$ for all $\alpha, \beta, \ldots$, is then evaluated:

$$f_0^{DW} = \binom{\tilde{N}}{2}\left(\frac{1}{\tilde{N}}\right)^4 \propto \left(\frac{1}{\tilde{N}}\right)^2 \text{ for } \tilde{N} \gg 1. \tag{6.3}$$

This means that for $\tilde{N} > 3$, the energy of the junction decreases with the order $\tilde{N}$ and approaches 0 for large $\tilde{N}$. This must be termed "unphysical," since junctions between objects lose their penalty, and the system would return to the disordered state. The double-obstacle potential is introduced [12] to remedy this problem:

$$f_0^{DO} = \sum_{\alpha=1...\tilde{N}, \beta=\alpha+1...\tilde{N}} |\phi_\alpha\phi_\beta|. \tag{6.4}$$

This has the same topology as the double-well potential (see Fig. 2.2) but a maximum power of two. We calculate the maximum potential of the junction f_m^{DO}:

$$f_0^{DO} = \binom{\tilde{N}}{2}(\frac{1}{\tilde{N}})^2 \propto 1 \ \ \text{for} \ \ \tilde{N} \gg 1. \tag{6.5}$$

This means that the energy of the junction increases with the order $\tilde{N}$ and approaches a constant for large $\tilde{N}$, as it should. The main drawback of this potential is the non-analytical form with the absolute signs. But this is technical...

6.3 Multi-Phase-Field Equation

The most important feature of the multi-phase-field theory is the consideration of a sum constraint:

$$\sum_{\alpha} \phi_{\alpha} = 1. \tag{6.6}$$

The system must be closed in itself so that "holes" do not arise, and this has important consequences for further model development. The phase fields within a junction cannot be varied independently; $\frac{\delta \phi_{\beta}}{\delta \phi_{\alpha}} \neq 0$. Starting with a relaxation ansatz (2.6), we must apply the chain rule:

$$\frac{\partial \phi_{\alpha}}{\partial t} \propto - \left[\frac{\delta}{\delta \phi_{\alpha}} + \sum_{\beta \neq \alpha} \frac{\partial \phi_{\beta}}{\partial \phi_{\alpha}} \frac{\delta}{\delta \phi_{\beta}} \right] F. \tag{6.7}$$

The variation $\frac{\partial \phi_{\beta}}{\partial \phi_{\alpha}}$, however, is generally unknown. Therefore, so-called "interface fields" $\psi_{\alpha\beta}$ are introduced [12]:

$$\psi_{\alpha\beta} = \left[\frac{\delta}{\delta \phi_{\alpha}} - \frac{\delta}{\delta \phi_{\beta}} \right] F. \tag{6.8}$$

These have the important property of automatically conserving the sum constraint (6.6) (for details see [3, 12]). With the phase-field mobility $M_{\alpha\beta}^{\phi}$ defined for a pair of phases, we reformulate (6.7):

$$\frac{\partial \phi_{\alpha}}{\partial t} = - \sum_{\beta} \frac{M_{\alpha\beta}^{\phi}}{\tilde{N}} \psi_{\alpha\beta} = - \sum_{\beta} \frac{M_{\alpha\beta}^{\phi}}{\tilde{N}} \left[\frac{\delta}{\delta \phi_{\alpha}} - \frac{\delta}{\delta \phi_{\beta}} \right] F. \tag{6.9}$$

Now we have consequently decoupled a multi-phase problem into pairwise contributions. We end this lecture with two remarks, or warnings:

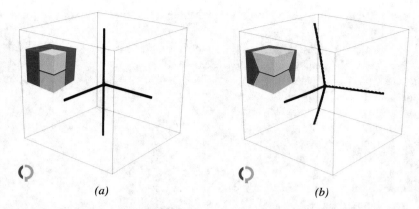

Fig. 6.2 Force balance of a quadruple junction between four grains with isotropic interface energy.
(**a**) The benchmark test starts with a simple configuration of rectangular grains. The triple lines
between three phases are displayed. The inset shows the phase configuration in color coding. (**b**) By
reduction of interface energy, the system relaxes to the final configuration with equal angles of 109°
between the triple lines and a symmetric grain structure

- As elegant as it looks, it is tedious to implement in phase-field software: since F
  is itself expanded in pairs, there is a third loop over phases $\gamma \neq \alpha \neq \beta$.
- As tedious it is to implement, it is nonetheless crucial if you are considering a
  real materials problem. Taking any shortcuts to avoid the general implementation
  will lead to a deadlock regarding the solution of the materials problem.

A kind of benchmark for the equilibrium configuration of junctions is displayed
in Fig. 6.2 where a brick-like initial structure relaxes into the equilibrium configuration [3, 4]. Note: This equilibrium configuration does not fit nicely into a rectangular
box. Therefore one has to employ some tricks regarding boundary conditions and
keeping the center within the box, it has the tendency to drift out, of course to
reduce interface energy. Such important details for practical applications, however,
are seldom discussed in close detail in the literature.

6.4   Exercise

Exercise
Prove that the form (6.8) for the interface field automatically conserves the
sum constraint (6.6) by adding a Lagrange multiplier $\lambda \left\{ \sum_\alpha \phi_\alpha - 1 \right\}$ to the
free energy F, see [12].

Example: Anisotropic Grain Growth

A multi-phase-field 3D grain growth model is formulated in [10] to simulate grain growth with anisotropic interface energy. The interface energy is defined as a function of the inclination angle of the boundary plane. It is found that the resulting grain-growth kinetics are strongly influenced by the anisotropy of the interface energy. This results in a slower growth rate and a distinct morphological evolution, as indicated by the presence of more cubic grains and a more significant number of triple-junction angles between 90° and 180°, as illustrated in Fig. 6.3. Additionally, the model closely matches various experimental results for NaCl and MgO polycrystalline minerals, where the distribution of grain-boundary planes peaks at low-index {100}-type boundaries.

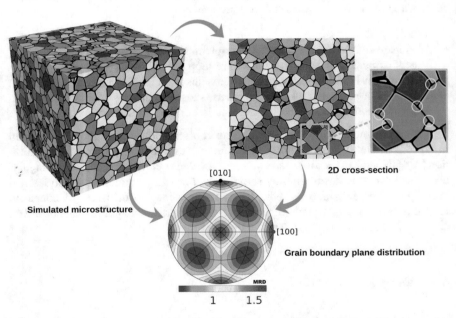

Fig. 6.3 Three dimensional simulation of anisotropic grain growth in 3D [10]. A 2D cross-section and the bow up of an insert shows clearly the deviation of the angles of interfaces at junctions from the isotropic angle expected for isotropic grain growth. Also the structure shows a pronounced cubic texture

Further Reading

- A multi-phase-field model without the sum constraint, but with a consistent treatment of the bulk-energy part, is presented in [8, 9].
- Force balance at junctions [3, 14].

References

1. R. Ahluwalia, T. Lookman, A. Saxena, Dynamic strain loading of cubic to tetragonal martensites. Acta Mater. **54**(8), 2109–2120 (2006). https://doi.org/10.1016/j.actamat.2005.12.040

2. D.N. Fan, L.Q. Chen, Computer simulation of grain growth using a continuum field model. Acta Mater. **45**, 611–622 (1997)

3. W. Guo, R. Spatschek, I. Steinbach, An analytical study of the static state of multi-junctions in a multi-phase field model. Physica D **240**(4–5), 382–388 (2011). https://doi.org/10.1016/j.physd.2010.09.014

4. W. Guo, I. Steinbach, Multiphase field study of the equilibrium state of multijunctions. Int. J. Mater. Res. **101**, 480–485 (2010). https://doi.org/10.3139/146.110298

5. R. Kobayashi, J.A. Warren, W.C. Carter, A continuum model of grain boundaries. Physica D **140** (1–2), 141–150 (2000). ISSN: 0167-2789. https://doi.org/10.1016/S0167-2789(00)00023-3

6. B. Korbuly, et al., Orientation-field models for polycrystalline solidification: grain coarsening and complex growth forms. J. Crystal Growth **457**, 32–37 (2017). 5th European Conference for Crystal Growth (ECCG), Bologna, Italy, Sep 2015, pages 32–37. ISSN: 0022-0248. https://doi.org/10.1016/j.jcrysgro.2016.06.040

7. E. Miyoshi, T. Takaki, Extended higher-order multi-phase-field model for three-dimensional anisotropic-grain-growth simulations. Comput. Mater. Sci. **120**, 77–83 (2016). ISSN: 0927-0256. https://doi.org/10.1016/j.commatsci.2016.04.014

8. N. Moelans, A quantitative and thermodynamically consistent phase-field interpolation function for multi-phase systems. Acta Mater. **59**, 1077–1086 (2011)

9. N. Moelans, F. Wendler, B. Nestler, Comparative study of two phase-field models for grain growth. Comput. Mater. Sci. **46**, 479–490 (2009)

10. H. Salama, et al., Role of inclination dependence of grain boundary energy on the microstructure evolution during grain growth. Acta Mater. **188**, 641–651 (2020). https://doi.org/10.1016/j.actamat.2020.02.043

11. O. Shchyglo, U. Salman, A. Finel, Martensitic phase transformations in Ni–Ti-based shape memory alloys: the Landau theory. Acta Mater. **60**(19), 6784–6792 (2012). https://doi.org/10.1016/j.actamat.2012.08.056

12. I. Steinbach, F. Pezzola, A generalized field method for multiphase transformations using interface fields. Physica D **134**, 385–393 (1999). https://doi.org/10.1016/S0167-2789(99)00129-3

13. I. Steinbach, et al., A phase field concept for multiphase systems. Physica D **94**(3), 135–147 (1996). https://doi.org/10.1016/0167-2789(95)00298-7

14. T. Young, An essay on the cohesion of fluids. Philos. Trans. R. Soc. **95**, 65–87 (1805). https://doi.org/10.1098/rstl.1805.0005

Chapter 7
Stress–Strain and Fluid Flow

7.1 Coupling to Stress and Strain

In this last lecture, before we change gears to the "quantum-phase-field" model, we want to (almost) complete the picture of the effects to be considered in real materials. "Almost," because we discuss only elasticity and fluid flow; we only touch on plasticity, and we neglect electric charges, magnetic coupling, and others. As stated in Chap. 1, a wide range of applications of phase-field modelling and simulations lies in this field. The models applied there, however, are complementary in concept with what we will discuss now: coupling of phase evolution with elastic distortion.

It is reported that Armen Khachaturyan developed his branch of phase fields, the so-called "time-dependent Ginsburg–Landau theory," to predict the equilibrium shape of a martensite needle in parent austenite. The problem relates to the old Eshelby problem of an inclusion in a matrix with finite transformation strain [9]. We may also include deviatoric strain if the crystal lattice changes from BCC to FCC. This was used to find the solution to the problem of the equilibrium shape of a crystal in its melt using the anisotropy of the interface energy in Chap. 3. Practically speaking, and this is often reported in the literature (to which I give no reference here), you cut out a piece of material from a crystal, transform it to a different phase with volumetric and deviatoric distortion with respect to the original crystal, and put it back. Continuity demands that both the parent and child crystal have to deform elastically to form a common body (see Fig. 7.1).

In the phase-field approach, this is an easy exercise: you simply have to add the elastic bulk free energy of the individual phases to our previous models and decide on a good way to handle the elastic contribution within the interface. The elastic bulk free energy f^{elast} is (and all phase-field models agree with classical models from continuum mechanics on this point):

© The Author(s) 2023
I. Steinbach, H. Salama, *Lectures on Phase Field*,
https://doi.org/10.1007/978-3-031-21171-3_7

Fig. 7.1 Scheme of the
elastic equilibrium between
a transformed inclusion and
the matrix

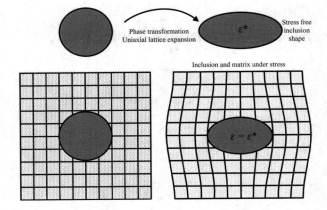

$$f^{\text{elast}} = \frac{1}{2} \left\{ \sum_{\alpha=1}^{\tilde{N}} \phi_\alpha (\epsilon_\alpha^{ij} - \epsilon^{*ij}_\alpha) C_\alpha^{ijkl} (\epsilon_\alpha^{kl} - \epsilon^{*kl}_\alpha) \right\}, \qquad (7.1)$$

where ϵ_α^{ij} is the total strain in phase α, ϵ^{*ij}_α is the eigenstrain (or transformation free strain), and C_α^{ijkl} is the elasticity matrix. We use the sum convention for double indices.

In general, ϵ^{*ij}_α and C_α^{ijkl} are concentration- and temperature-dependent quantities, which gives rise to intriguing effects of chemo-mechanical coupling (see [16] and references therein). We leave this for further reading. The ansatz (7.1) is a direct extension of the original multi-phase model for diffusive phase transformations, as the total elastic energy is a linear summation of the elastic energies of the individual phases weighted by the phase densities ϕ_α. It is mostly assumed that the equilibrium is instantaneous. One derives the mechanical equilibrium equation, as we do with all relevant transport equations, as a functional derivative from the free-energy functional

$$\vec{0} = \nabla\sigma = \nabla \frac{\delta}{\delta\epsilon} F \qquad (7.2)$$

in vector notation, where the stress σ and strain ϵ are rank-two tensors. The difficulty now is to define these tensors within the interface, because the definition of the elastic free energy density (7.1) uses different strains ϵ_α. In all bulk phases, the stress is of course $\sigma = \epsilon C$, but also the elasticity C is only defined for each phase. This is a classical problem in solid mechanics: to define an effective material from a mixture of materials, so-called mathematical homogenization. To be able to solve the mechanical equilibrium equation (7.2), one first has to define the effective mechanical properties of the interface, the effective elasticity matrix C, and also the effective transformation strain ϵ^*. The latter is commonly taken as a weighted average $\epsilon^* = \sum_\alpha \phi_\alpha \epsilon_\alpha^*$. In the phase-field literature, the most-used homogenization for the total strain tensor ϵ is the so-called Voigt–Taylor model, which assumes homogeneous strain in all phases $\epsilon = \epsilon_\alpha = \epsilon_\beta$. Then, however, the stress has to be discontinuous in general. The opposite, the Reuss–Sachs model, takes the stress as continuous between the phases. Since the stress is the equivalent to a chemical potential in a mechanical system, one may take this assumption as an analogue to equal chemical potential in the interface (see also the discussion in [20]).

A combination of both models, taking the strain as continuous in the tangential direction and the stress as continuous in the normal direction to the interface, is proposed in [8]. The most advanced model is the so-called rank-one convexification applying a jump condition (see [15, 17]). We leave this topic to further reading. Only one important consequence shall be detailed here: the elastic driving force for phase transformations Δg^{elast} for a phase-transformation under mechanical load, either external or by internal stresses. For the Voigt–Taylor model, we find for the dual α–β interface (the general form is defined as a superposition between dual driving forces in the multi-phase-field approach):

$$\Delta g_{\alpha\beta}^{\text{elast}} = \left(\frac{\delta}{\delta\phi_\alpha} - \frac{\delta}{\delta\phi_\beta} \right) F$$

$$= (\epsilon_\alpha^* - \epsilon_\beta^*)(\epsilon - \epsilon^*)C + \frac{1}{2}(\epsilon - \epsilon^*)^2(C_\alpha - C_\beta). \tag{7.3}$$

For Reuss–Sachs, we have:

$$\Delta g^{\text{elast}} = (\epsilon - \epsilon^*)C(\epsilon_\alpha^* - \epsilon_\beta^*)C \left[(\epsilon_\alpha^* - \epsilon_\beta^*) - \frac{1}{2}\left(\frac{1}{C_\alpha} - \frac{1}{C_\alpha} \right)C(\epsilon - \epsilon^*) \right]. \tag{7.4}$$

We see in both cases, that the elastic driving force for phase transformations vanishes for a homogeneous material if $\epsilon_\alpha^* = \epsilon_\beta^*$ AND $C_\alpha = C_\beta$. Also in both cases we have two contributions: one related to the difference in the eigenstrains, the other related to the difference in the elasticity matrices. And of course, the elastic driving force vanishes if the total strain matches the eigenstrains, i.e., there is no stress in the interface region. Higher-order schemes considering a jump condition between stress and displacement at the interface [8, 15, 17] are more involved, but they follow the same principles.

Another remark relates to the solution of the mechanical equilibrium equation (7.2). This can be achieved by any appropriate numerical approach using a finite-element scheme in real space. In phase-field models, the Fourier method is very popular because its solution for homogeneous systems, i.e., $C = C_\alpha = C_\beta$, is computationally very cheap. If the material cannot be approximated by homogeneous elasticity, one first solves a homogeneous problem and then corrects the solution for the inhomogeneity in an iterative scheme. The difference between the elasticity coefficients is called "contrast." In the case of a low contrast, i.e., if the elasticity in all phases is similar, these schemes are quite effective. For high contrast, e.g., a pore filled with gas or liquid in a solid metal, one needs more elaborate schemes [12].

One also has to consider that the Fourier transformation relies on periodic boundary conditions. In any case, however, one should not confuse the theoretical setup of the phase-field model with a special solution procedure, as important as this is for application.

The final comment regards plasticity. There may be an outer load that exceeds the yield point of the material. Then, elastic strain is limited, and plastic strain sets in. This is associated with dislocation activity and the generation of another contribution to the free energy of the system: stored plastic energy related to dislocations. Such stored plastic energy may lead to phenomena such as "recrystallization." This can be treated similarly to a phase transformation, and it is driven by the reduction of plastic energy. This is "easy" from a phase-field perspective, but it is difficult in terms of handling plastic energy at the continuum scale. It is even more difficult to design a model accounting for how moving interfaces interact with dislocations. Some applications are suggested for further reading. Additionally, internal stresses, which are caused by transformation strain between different phases, may exceed the yield point of the material, in particular at elevated temperatures. The stress state of the interface, and thus the driving force for phase transformation, is then limited by plastic relaxation. Therefore, plasticity will generate new driving forces on the one hand, and it will reduce driving forces on the other hand. This is a wide area for future research, but it is associated with special challenges regarding (i) model formulation and (ii) numerical solution.

Example—Martensitic Transformation
A body-centered-tetragonal martensite structure is formed by rapidly quenching from the face-centered-cubic austenite phase. When a material undergoes a martensitic transformation, the crystalline unit cell undergoes a shape change. The final microstructure of martensite is strongly influenced by elastic energy generated during the change of the crystal structure. Accommodation of the elastic energy is often achieved by forming a multi-variant domain. In addition, plastic deformation occurs in the austenite matrix and the growing martensite phase, and these are called self-accommodation and plastic accommodation, respectively [5, 23]. Thus, both elastic and plastic effects play important roles in martensitic transformations.

The Simulation
A 3D multi-phase-field model was used to model martensite microstructure formation in low-carbon steel. In this example (Fig. 7.2), an elasto-plastic approach is employed with a full set of 24 Kurdjumov–Sachs symmetry variants of martensite with real transformation strains [13]. A finite strain

(continued)

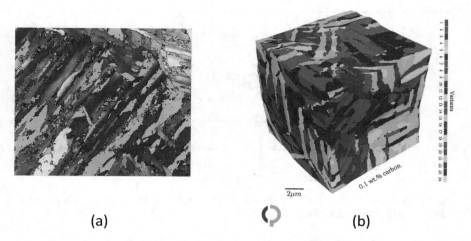

Fig. 7.2 Lath martensite microstructures: (**a**) electron backscatter diffraction map of martensite in a sample containing 0.1 wt.% carbon [18]; and (**b**) simulated martensite microstructure in low-carbon steel consisting of 24 Kurdjumov–Sachs symmetry variants, indicated by color coding

framework is applied in connection to a phenomenological crystal-plasticity model. Both methods are fully coupled within the **OpenPhase** library.

7.2 Coupling to Fluid Flow

In previous lectures, we have discussed the coupling of the phase field with temperature, solute, and elastic distortion. In all these cases, the driving force for a phase transformation $\Delta g_{\alpha\beta} = \left(\frac{\delta}{\delta\phi_\alpha} - \frac{\delta}{\delta\phi_\beta} \right) F$ is directly dependent on temperature, composition, or stress. We consider this coupling as "strong." In the present subsection, we consider fluid flow, or melt flow when considering a solidification process. The coupling to phase evolution may be considered as "weak," since the direct effect of flow on a phase transformation, the Clausius–Clapeyron effect, is generally very weak. The Clausius–Clapeyron effect describes the boiling temperature of a liquid dependent on the pressure in the system: the boiling temperature of water is significantly reduced on high mountains. The melting temperature of a metal, however, is hardly affected by the pressure of a streaming melt, and we therefore simply neglect it. There is, however, a strong effect of melt flow on dendrite morphologies in solidification. This is because on the one hand, melt flow significantly affects the transport of solute and heat in the melt; on the other hand, solidification significantly affects the viscosity of the material.

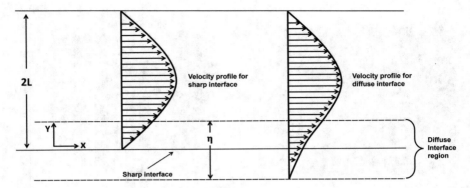

Fig. 7.3 Schematic of channel flow in which the width of the channel and the width of the interface are of the same order of magnitude

A solid dendrite can be seen as a rigid body. If it is attached to the mold of the casting, it forms a rigid barrier for flow; if it is transported with the melt as an equiaxed crystal, it will affect the effective viscosity of the two-phase system (solid and melt). If the solid fraction is low, a metallic melt has a viscosity like that of water; if the solid fraction exceeds 30%, it will behave like honey with precipitated sugar crystals; above 50%, any melt flow will stop.

This mutual interaction leads to intriguing phenomena, which we will leave for further reading. Here, we will elaborate on one effect that is crucial for the interaction of a fluid with a solid when the interface between liquid and solid is diffuse at a mesoscopic scale. At the microscopic scale, we definitely accept a so-called "no-slip" condition: the flow velocity decays to 0 monotonously in the direction normal to the interface. Figure 7.3 schematically shows the condition of a channel flow in which the width of the channel and the width of the interface are of the same order of magnitude. One side of the channel is treated as a "wall," i.e., a sharp interface with a well-defined no-slip condition; the other side is treated as diffuse in the context of phase-field theory. How do we realize an analogue to the no-slip condition within a diffuse boundary? As in the first part of this lecture, we treat the interface as an effective material, and in the present case as a porous medium: the flow will penetrate the interface, but the permeability of the interface will be a function of the phase field.

The corresponding fluid-flow equation for the fluid velocity $\vec{u}$ in a mixed domain—solid and liquid—is readily written down (for simplicity, without moving solids, i.e., the solid velocity is set to 0; $\phi = \phi_{\text{liquid}}$, density ρ set to 1):

$$\frac{\partial}{\partial t}\phi\vec{u} + \vec{\nabla}\phi\vec{u}\vec{u} = -\phi\vec{\nabla}P + \vec{\nabla}\left(\nu\nabla\phi\vec{u}\right) - h^*X_l. \tag{7.5}$$

The important part here is the last term: friction within the diffuse interface X_l. This has been worked out for the planar interface with a friction coefficient h^*.[1] Its numerical value is given as $h^* = 2.757$ [6]. Other models defining the no-slip condition as a function of the phase field have been investigated [3, 4, 21] and may be investigated in the future. A consistent investigation for curved interfaces (concave or convex) is still missing. In any case, we will require the "sharp-interface solution" to be met outside of the interface, i.e., that we meet the sharp-interface model in the bulk. In the case of the channel flow, the sharp-interface model predicts a parabolic velocity profile in the channel: Hagen–Poiseuille law. The maximum velocity is a sensitive indicator of whether the diffuse interface correctly emulates the sharp interface; it is a kind of "thin-interface limit," i.e., we match a sharp-interface solution in the bulk liquid but deviate systematically within the thin interface.

7.3 Exercises

Exercise
Derive the expressions for the elastic driving forces (7.3) and (7.4) for the Voigt–Taylor and Reuss–Sachs limit, respectively.

Example: Dendritic Solidification Interacting with Shear Flow of the Melt
The setup (Fig. 7.4) represents a thin Mg-Al melt channel between two rigid walls. We see nucleation and growth of α-Mg dendrites. At the same time, shear flow is introduced to the melt via the motion of the rigid walls. Full integration of the inertial and friction forces acting on the solid dendrites and the melt results in the dendrites moving with the melt. In this example, the fluid-flow problem is solved using the lattice Boltzmann method, while the system morphology and its evolution is described by the phase-field method. Both methods are fully coupled within the **OpenPhase** library, allowing the study of arbitrarily complex geometries. See [14, 22].

---

[1] h^* is named the "Hermann-Joseph" constant. Hermann-Joseph Diepers was one of my early collaborators. He passed away from cancer before finishing his PhD. His memory lasts.

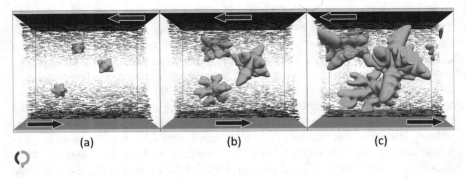

Fig. 7.4 Flow simulation of Mg–Al alloy solidification

Further Reading

- Chemo-mechanical coupling [10, 16].
- Hadamard jump [15, 17].
- Recrystallization and rafting under high-temperature creep [1, 2, 7, 11].
- Dendritic growth with buoyancy [19, 22].

References

1. M. Ali, et al., 45-degree rafting in Ni-based superalloys: a combined phase-field and strain gradient crystal plasticity study. Int. J. Plast. **128**, 102659 (2020). https://doi.org/10.1016/j.ijplas.2020.102659
2. M. Ali, J.V. Görler, I. Steinbach, Role of coherency loss on rafting behavior of Ni-based superalloys. Comput. Mater. Sci. **171**, 109279 (2020). https://doi.org/10.1016/j.commatsci.2019.109279
3. D.M. Anderson, G.B. McFadden, A.A. Wheeler, A phase-field model of solidification with convection. Physica D **135**, 175–194 (2000)
4. D.M. Anderson, G.B. McFadden, A.A. Wheeler, A phase-field model of solidification with convection: sharp-interface asymptotics. Physica D **151**, 305–331 (2001)
5. G.R. Barsh, et al., A new view on martensitic transformations. Scripta Metallurgica **21**(9), 1257–1262 (1987)
6. C. Beckermann, et al., Modeling melt convection in phase-field simulations of solidification. J. Comput. Phys. **154**, 468–496 (1999)
7. S. Chakraborty, et al., Investigating the origin of cube texture during static recrystallization of FCC metals: a full field crystal plasticity-phase field study. arXiv: 2006.06475 [cond-mat.mtrl-sci] (2020)
8. A. Durga, P. Wollants, N. Moelans, A quantitative phase-field model for two-phase elastically inhomogeneous systems. Comput. Materi. Sci. **99**, 81–95 (2015). https://doi.org/10.1016/j.commatsci.2014.11.057
9. J.D. Eshelby, The elastic field outside an ellipsoidal inclusion. Proc. R. Soc. Lond. A **252**, 561–569 (1959). https://doi.org/10.1098/rspa.1959.0173

10. J. Görler, et al., Gamma-channel stabilization mechanism in Ni-base superalloys. Philos. Mag. Lett. **95**(11), 519–525 (2015). https://doi.org/10.1080/09500839.2015.1109716

11. J. Hiebeler, et al., Modelling of flow behaviour and dynamic recrystallization during hot deformation of MS-W 1200 using the phase field framework, in *MATEC Web of Conferences*. EDP Sciences, vol. 80 (2016), p. 01003. https://doi.org/10.1051/matecconf/20168001003

12. S.Y. Hu, L.Q. Chen, A phase-field model for evolving microstructures with strong elastic inhomogeneity. Acta Mater. **49**(11), 1879–1890 (2001). https://doi.org/10.1016/s1359-6454(01)00118-5

13. G. Kurdjumow, G. Sachs, Über den mechanismus der stahlhärtung. Z. Phys. **64**(5–6), 325–343 (1930)

14. D. Medvedev, F. Varnik, I. Steinbach, Simulating mobile dendrites in a flow. Proc. Comput. Sci. **18**, 2512–2520 (2013)

15. J. Mosler, O. Shchyglo, H. Montazer Hojjat, A novel homogenization method for phase field approaches based on partial rank-one relaxation. J. Mech. Phys. Solids **68**, 251–266 (2014). https://doi.org/10.1016/j.jmps.2014.04.002

16. J. Park, et al., First evidence for mechanism of inverse ripening from in-situ TEM and phase-field study of δ' precipitation in an Al–Li alloy. Sci. Rep. **9**, 3981 (2019). https://doi.org/10.1038/s41598-019-40685-5

17. D. Schneider, et al., Phase-field elasticity model based on mechanical jump conditions. Comput. Mech. **55**(5), 887–901 (2015). https://doi.org/10.1007/s00466-015-1141-6

18. O. Shchyglo, et al., Phase-field simulation of martensite microstructure in low-carbon steel. Acta Mater. **175**, 415–425 (2019)

19. I. Steinbach, Pattern formation in constrained dendritic growth with solutal buoyancy. Acta Mater. **57**, 2640–2645 (2009). https://doi.org/10.1016/j.actamat.2009.02.004

20. I. Steinbach, M. Apel, Multi phase field model for solid state transformation with elastic strain. Physica D **217**, 153–160 (2006). https://doi.org/10.1016/j.physd.2006.04.001

21. A. Subhedar, I. Steinbach, F. Varnik, Modeling the flow in diffuse interface methods of solidification. Phys. Rev. E **92**(2), 023303 (2015). https://doi.org/10.1103/PhysRevE.92.023303

22. M. Tegeler, et al., Effect of microstructure during dendritic solidification on melt flow: a phase-field lattice-Boltzmann study, in *Proceedings of the 6th Decennial International Conference on Solidification Processing* (2017)

23. A. Yamanaka, T. Takaki, Y. Tomita, Elastoplastic phase-field simulation of self-and plastic accommodations in cubic → tetragonal martensitic transformation. Mater. Sci. Eng. A **491**(1–2), 378–384 (2008)

Chapter 8
Quantum Phase Field

8.1 Introverted Picture of Mass and Space

How to Start? There are so many aspects coming together to form a consistent theory, culminating in phase-field theory. From the physical point of view, this includes thermodynamics, wave mechanics, interface-driven phenomena and capillarity, and the kinetics of phase transformations. From the practical point of view, we have the numerics of the solutions of partial differential equations, and multi-physics problems. The main applications of phase-field theory to date have been engineering problems. From the mathematical point of view, we derive a set of partial differential equations using variational principles from a governing functional that has a gradient contribution acting on the field. In quantum-mechanical language, this gradient is a momentum operator, and the field is a quantum field with a discrete spectrum. If this sounds strange, then be reassured that, as before, we will go through this step by step.

Let us start from the conceptual point of view. We discuss a free-boundary problem. The free boundary is the inner boundary between two or more domains, which are characterized by phase fields. In classical phase-field theory, we call the boundary an interface. The boundary is termed "free" because it is not fixed by some external condition; it can evolve in time (see Chap. 1). This inner boundary is different from outer boundaries, i.e., the boundaries of the calculation domain, which are fixed. The inner boundary between phases evolves in space and time. This evolution is closely coupled to transport phenomena in the domains: transport of temperature, solute, and momentum, as well as displacement. These phenomena have been discussed in the previous lectures.

The object of interest in phase-field theory is the inner boundary. This is not something that we specify; it is something that will come out: it "emerges." It shall be independent of the settings of the outer boundary in any system; if there is any outer boundary at all.

© The Author(s) 2023
I. Steinbach, H. Salama, *Lectures on Phase Field*,
https://doi.org/10.1007/978-3-031-21171-3_8

Fig. 8.1 The billiard table as a representation of the extroverted view of space in traditional physics: objects are placed in space. The network of introverted spaces (yellow lines) connecting the balls reverts the view to a closed network system without outer boundaries (© www.winsport. de with permission)

Boundaries bound "spaces". We may distinguish between "introverted" and "extroverted" spaces. The latter means the space around an object. The extroverted space is a kind of background, existing with or without the object. This is the traditional view of space. Objects are placed in space. As an illustration see Fig. 8.1. The green of the billiard table represents space. The objects, billiard balls, are placed in this space. They interact according to Newton's laws, momentum and energy conservation. In a real billiard game the player uses the cushions to mirror space. We may, hypothetically, push the cushions to infinity, then our play ground will be infinitely large, but balls would never come back. We may postulate periodic or "Moebius" boundary conditions. But all this becomes irrelevant if we revert the picture to an introverted space. Introverted space lives inside the objects: it is bounded by the objects. These introverted spaces are sketched by the yellow lines in Fig. 8.1. The relative position of all balls is uniquely determined by the network of inner spaces. An absolute outer space is not needed to characterize their interrelation.

In the physical world, which is composed of particles and space, the only objects that could serve as a bound of an introverted space are elementary particles. So, if particles (zero-dimensional objects) are the boundaries, space is the 1-dimensional line-distance between particles. There is no other choice.

If there are many particles, there will be many spaces. In the quantum-phase-field concept, each space is associated with a phase field ϕ_α. From the number of elementary particles within the observable universe N, a maximum of N combinatorial 2 spaces is a number of the order of 10^{120}. Space will also be taken as "substantial," i.e., it is not just "empty": it has an energy, a negative energy related to vacuum fluctuations, as shown in Sect. 8.3. The 3D "space of cognition," the space of our daily experience, will be reconstructed later, in Sect. 8.4. We will proceed like this: Define a system of quantum phase fields, closed in itself, without having outer boundaries.

8.2   Formal Definition of Quantum Phase Fields

We define (and you have all the background from the previous lectures) a set of a large number of phases ϕ_α. The phases form a system that is closed in itself, forming a "universe." It is a system without outer boundaries. We do this in analogy to the multi-phase-field system from Chap. 6:

$$\sum_\alpha \phi_\alpha = 1. \tag{8.1}$$

Here, ϕ_α is simply termed the "phase" since up to now no "space" has been defined. Each phase ϕ_α is an element of the universe that is different from all other elements ϕ_β. Therefore, an interface between these phases is formed: a fermionic particle, as we will see. The phases incorporate the basic elements of physical matter, "mass" and "space." We will associate them with the conserved quantity: energy H. The previous statement that "the phases incorporate mass and space" is important: the phases (our phase fields) ϕ_α are not defined *in* space, as in classical or quantum field theories, but rather they *define* space! Massive elementary particles are not "placed into space," they "connect spaces." We will derive this step by step.

The energy functional of the system in quantum mechanics $\hat{H}$ is an operator (as indicated by the hat). This is acting on the wave function of the system $|w\rangle$. In our case, the system—though this sounds quite highbrow—is the universe; $|w\rangle$ is the wave function of the universe. The only ingredient of the theory, up to now, is energy $H = \langle w|\hat{H}|w\rangle$ as the expectation value of $\hat{H}$. Energy is conserved (first law of thermodynamics).[1] Furthermore, we will postulate that $H = 0$, i.e., there is no net energy: all energetic states, positive and negative, have to sum up to 0. The argument for this is simple: there is no evidence regarding where a finite energy should come from (see also the Wheeler–DeWitt theory [7]).

We will allow changes $d\hat{H}$, that the zero-energy state, the state of "nothing," separates into positive and negative energetic states, the state of "something." This may be related to the Big Bang as the origin of our universe, if you will. Changes are related to a non-conserved quantity, entropy: the second law of thermodynamics. We will expand $\hat{H}$, in the changes $d\hat{H}$ with respect to the phases ϕ_α. $\hat{H}$ thereby is itself a function of all fields $\{\phi_\beta\}$, $\hat{H} = \hat{H}(\{\phi_\beta\})$, and, as a reminder, all fields are connected by the sum constraint (8.1):

$$\hat{H} = \sum_\alpha \int_0^1 d\phi_\alpha \frac{\partial \hat{H}(\{\phi_\beta\})}{\partial \phi_\alpha}. \tag{8.2}$$

---

[1] Note that the constraint of energy conservation is in contradiction to general relativity [8], where energy is not conserved! We will accept this.

The integral runs over the definition range of the phase fields from 0 to 1, meaning **yes or no, existing or not-existing**, and we allow the fields to vary between these bounds, i.e., that they are diffuse, as is usual in phase-field theory.

The phases will be used to indicate different states of energy. This means that they will be used to derive relations between the different states of energy. We will allow a **BIG** number of fields connecting a **BIG** number of different states of energy, both positive and negative. An important factor in this regard is that positive and negative states are not just "mirrored" states like matter and anti-matter in traditional physics: they are topologically different, as we will discuss in closer detail when we discuss the ordering scheme of space and mass in Sect. 8.4. For now we simply state, that there are negative energy states E_α associated to a phase α, and positive states $U_{\alpha\beta}$ associated to junctions between phases α and β.

There is no fundamental space. We repeat the argument: where should it come from?

"Space" is introduced as a distance Ω_α, defined by the inverse of the negative state of energy $E_\alpha < 0$ of the field ϕ_α:

$$\Omega_\alpha = -\tilde{\alpha} \frac{hc}{48E_\alpha}, \tag{8.3}$$

where h is Planck's constant, c is the speed of light, and $\tilde{\alpha}$ is a positive and dimensionless coupling parameter to be determined. It will be proven self-consistently in Sect. 8.3 that this definition leads to a 1D line coordinate s_α specific to each phase, $\phi_\alpha = \phi_\alpha(s_\alpha)$. Having this line coordinate we can treat $\phi_\alpha(s_\alpha)$ as a classical phase field in 1D.

There is no proof that this is the only possible construction to define our universe, no rigorous argument against concepts of "parallel universes" or "multiverses," or 10- or 21-dimensional spaces with strings and branes embedded, as postulated by several researchers (see, e.g., [14] for 10 dimensions and [18] for 21 dimensions). You will find almost every number of dimensions for space–time postulated from 4 to 21, maybe even higher, since the original work of Kaluza and Klein [15, 16]. A whole discipline, called "mathematical physics," is searching for a "unitarian description of all physical phenomena" in multidimensional manifolds with appropriate geometrical measures.

We will approach this differently. The present theory introduces time and space as auxiliary coordinates: the coordinates are not "fundamental." It is shown that the theory, which is based on a set of principal statements, is consistent with these statements when formulated in these auxiliary coordinates. It is "self-consistent," and it resembles our cognition of the physical world.

Substituting $d\phi_\alpha = \frac{\partial \phi_\alpha}{\partial s_\alpha} ds_\alpha$ and introducing the forces $\hat{h}_\alpha = \frac{d\hat{H}}{ds_\alpha}$ yields

$$\hat{H} = \sum_{\alpha=1}^{N} \int_{-\infty}^{\infty} ds_\alpha \frac{\partial \phi_\alpha}{\partial s_\alpha} \frac{\partial \hat{H}(\{\phi_\beta\})}{\partial \phi_\alpha} = \sum_{\alpha=1}^{N} \int_{-\infty}^{\infty} ds_\alpha \hat{h}(\{\phi_\beta\}). \tag{8.4}$$

Fig. 8.2 Two doublons in a periodic setting. Each doublon is formed by a right-moving and a left-moving soliton. The velocity is proportional to the energy difference between adjacent doublons

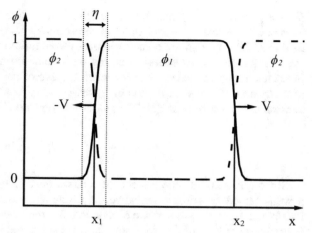

Space emerges, i.e., it is created by variations in the phases $d\phi_\alpha$: compare the "diffuseness of phase field." We relate the line coordinate s_α to the distance Ω_α, defined by the inverse of the energy quantum E_α (see (8.3)), by the integral

$$\int_{-\infty}^{+\infty} ds_\alpha \phi_\alpha = \Omega_\alpha = -\tilde{\alpha}\frac{hc}{48E_\alpha}. \tag{8.5}$$

This is easily be seen from the solution of $\phi_\alpha(s_\alpha)$, Fig. 8.2. Phases and space are complementary objects, defined by phase fields. In the present concept, the phase field *generates* space instead of *living in a space*. But we need to swallow a toad: space s_α is a 1D line coordinate. There is no evidence of a 3D space (besides our daily cognition, *which might be in error!*).

To construct the energy operator, or Hamiltonian $\hat{H}$, we use the standard form of the Ginzburg–Landau functional in 2D Minkowski notation:[2]

$$\hat{H} = \sum_{\alpha=1}^{N} \int_{-\infty}^{+\infty} ds\,\frac{4U\eta}{\pi^2}\left\{\left(\frac{\partial}{\partial s}\phi_\alpha\right)^2 - \frac{1}{c^2}\left(\frac{\partial}{\partial t}\phi_\alpha\right)^2 + \frac{\pi^2}{\eta^2}|\phi_\alpha(1-\phi_\alpha)|\right\}. \tag{8.6}$$

You will recognize the similarity with the free-energy model with a double-obstacle potential from Chap. 2. In addition to the gradient contribution in space, we include a gradient contribution in time. The time derivative accounts for dissipation, i.e., that in contrast to gradients in space, which have a positive energy penalty, gradients in time are favored energetically: the system shall evolve and not stagnate. We may treat this philosophically or simply state that this ansatz ensures relativistic invariance due to Lorentz contraction of the interface width (see [20] for details).

---

[2] We have removed the index α from space and time coordinates for readability.

U is a positive energy quantum to be associated with the positive rest-mass of elementary particles, the equivalent of the surface energy in the traditional phase-field model.[3] The gradient contributions of the Hamiltonian, Eq. (8.6), $\frac{\partial}{\partial s}$ and $\frac{1}{c}\frac{\partial}{\partial t}$, shall be understood as operators acting on a quantum mechanical wave function $|w\rangle$. This allows to evaluate the expectation value of the energy for an actual state of the system. The wave function in "quasi-static approximation" will be explicitly constructed below in Sect. 8.3. The phase-field equation is written down:

$$\tilde{\tau}\frac{\partial}{\partial t}\phi_\alpha = -\frac{\delta}{\delta\phi_\alpha}\int_0^{+\infty} dt \langle w|\hat{H}|w\rangle. \tag{8.7}$$

Since gradients of time are considered, we have to integrate over time to have a well-defined functional derivative, as discussed in Chap. 2. This equation has two parts: (i) a non-linear wave equation for the phase field; and (ii) a linear Schrödinger-type equation for quantum-mechanical excitations. This procedure is not new; it can be traced back to the so-called de Broglie–Bohm double-solution program [1, 2, 5, 6]. It is fully accepted as an alternative interpretation of quantum mechanics compared to the prevailing so-called "Copenhagen interpretation." For more explanation, see [17].

Now, we separate the expectation value of the energy functional (8.6) into three different contributions. These are distinguished by whether the differential operators $\frac{\partial}{\partial s}$ and $\frac{\partial}{\partial t}$ are applied to the wave function $|w\rangle$ or the field ϕ_α.

Applying the differential operators to the phase-field components and using the normalization of the wave function $\langle w|w\rangle = 1$ yields the force u_α $[\frac{J}{m}]$ related to the gradient of the fields α:

$$u_\alpha = \frac{4U\eta}{\pi^2}\left[\left(\frac{\partial\phi_\alpha}{\partial s}\right)^2 - \frac{1}{c^2}\left(\frac{\partial\phi_\alpha}{\partial t}\right)^2 + \frac{\pi^2}{\eta^2}|\phi_\alpha(1-\phi_\alpha)|\right]. \tag{8.8}$$

This contribution relates to the interface energy in the conventional 3 dimensional phase field application. In 1 dimensions it is a force quantum, related to gradients of the phase field or to values of the phase-field $0 < \phi_\alpha < 1$.

The mixed contribution, when one of the operators $\frac{\partial}{\partial s}$ and $\frac{\partial}{\partial t}$ is applied to the field ϕ and one to the wave function $|w\rangle$, describes the correlation between the field and the wave function. It shall be set to 0 in the quasi-static limit. In this limit, we keep the field static for the evaluation of the quantum-mechanical force. Then, we take this force for the determination of the time evolution of the field. A coupled solution has not been worked out to date:

$$0 = (1 - 2\phi_\alpha)\frac{4U\eta}{\pi^2}\left[\frac{\partial\phi_I}{\partial s}\langle w|\frac{\partial}{\partial s}|w\rangle - \frac{1}{c^2}\frac{\partial\phi_\alpha}{\partial t}\langle w|\frac{\partial}{\partial t}|w\rangle\right]. \tag{8.9}$$

---

[3] Note that the unit here is Joule, J, instead of $\frac{J}{m^2}$ for the surface energy in 3D, because we are living in a 1D space.

It is shown in [20] that this Eq. (8.9) is consistent with Newton's second law of acceleration. Finally, we apply the momentum operators $\frac{\partial}{\partial s}$ and $\frac{1}{c}\frac{\partial}{\partial t}$ to the wave function $|w\rangle$, which yields the force e_α $[\frac{J}{m}]$:

$$e_\alpha = \frac{4U\eta}{\pi^2}\phi_\alpha^2\langle w|\frac{\partial^2}{\partial s^2} - \frac{1}{c^2}\frac{\partial^2}{\partial t^2}|w\rangle. \tag{8.10}$$

This contribution applies to the bulk energy of the phase field $\phi_\alpha = 1$. We will explicitly evaluate this after the structure of the solutions of the fields is discussed. For a steadily moving field with velocity v, one transforms the phase-field equation into the moving frame traveling with this velocity $\frac{\partial}{\partial t} = v\frac{\partial}{\partial s}$. Inserting Eqs. (8.8)–(8.10) into (8.7), we find:

$$\tilde{\tau}\frac{\partial}{\partial t}\phi_\alpha = -\frac{\delta}{\delta\phi_\alpha}\int_0^{+\infty} dt\langle w|\hat{H}|w\rangle$$

$$= U\left[\eta\frac{\partial^2\phi_\alpha}{\partial s^2}\left(1 - \frac{v^2}{c^2}\right) + \frac{\pi^2}{\eta}\left(\phi_\alpha - \frac{1}{2}\right)\right] + m_{\phi_\alpha}\Delta e, \tag{8.11}$$

where: $\Delta e = e_\alpha - e_\beta$ is the difference in the volume force (the equivalent of the Gibbs free-energy difference) between two fields, Eq. (8.10); and m_{ϕ_α} is the appropriate coupling function. As a first result, we see that the interface width η contracts with velocity: $\eta_v = \eta\sqrt{1 - \frac{v^2}{c^2}}$, Lorentz contraction. It collapses for $v \to c$, and velocities $v > c$ are forbidden.

The structure of the field is well known to us from Chap. 2: the minimum solution of Eq. (8.11) is the "soliton." A periodic solution for two fields is displayed in Fig. 8.2. One field bounded by two solitonian waves, one right-moving and one left-moving, is called a "doublon." This is the basic element of physical space $\phi_\alpha = 1$, bounded by the junction to the other field, which will be interpreted as an elementary particle later.

8.3 Volume Energy of One Doublon

From the doublon solution in Fig. 8.2, we can see that the field forms a 1D box with fixed walls and size Ω_α for the field α. This is a standard exercise for a quantum problem: the particle in a box potential. The difference here is that the particle is massless, i.e., the dispersion relation is linear in momentum p instead of quadratic in p as for the massive particles. According to o Casimir [3], we have to compare quantum fluctuations in the box with discrete spectrum p and frequency $\omega_p = \frac{\pi cp}{2\Omega_l}$ to a continuous spectrum. This yields the negative energy E_α of the field α:

$$E_\alpha = \tilde{\alpha}\frac{hc}{4\Omega_\alpha}\left[\sum_{p=1}^{\infty} p - \int_1^{\infty} p\,dp\right] = -\tilde{\alpha}\frac{hc}{48\Omega_\alpha}, \tag{8.12}$$

where $\tilde{\alpha}$ is a positive, dimensionless coupling coefficient. I have used the Euler–Maclaurin formula in the limit $\epsilon \to 0$ after renormalization $p \to pe^{-\epsilon p}$. Since all parameters are positive, we see that "space" is accounted for by negative energy, scaling inversely proportional to the size of the doublon, which proves (8.3) for self-consistency. This energy scales like the energy of the gravitational field in Newtonian mechanics. Therefore, the force, as the derivative of the energy (8.12) with respect to space, can be associated with a gravitational attraction between the junctions between different doublons. The junctions can thus be interpreted as massive objects: elementary particles. They are associated with positive energy U [Eq. (8.8)] corresponding to the interface energy of a traditional phase field, while the bulk energy of the doublon is negative. In contrast to Newtonian mechanics and in agreement with general relativity (see "gravitational waves" [9]), the attraction is a wave phenomenon with time-dependent action. The coefficient $\tilde{\alpha}$ can be determined from the measured gravitational constant on Earth. Applying the theory to all masses of the visible universe gives a prediction for repulsive gravitation at ultra-long distances and an explanation of the expansion of the visible universe [20, 21].

8.4 Multidimensional Interpretation

As stated at the beginning, the present concept has no fundamental space. The distance Ω_α is intrinsic to one individual doublon α. Different doublons are connected in junctions due to the sum constraint (8.1): a multiple junction in the multi-phase-field terminology defines an elementary particle in the quantum-phase-field concept. The position of one particle related to an individual component of the field is determined by the steep gradient $\frac{\partial \phi_\alpha}{\partial s}$. In a multiple junction, where many fields intersect, the junction couples many different directions related to the different doublons. We shall consider the junction (or particle) as a "very small volume," in physics terminology, this is a zero-dimensional object. Since half-sided solitons are spinor-type objects belonging to the 3D SU(2) symmetry group, we may order the incoming and outgoing doublons of a junction in 3D Euclidean space. We will call this space the "space of cognition," since our cognition orders all physical objects in 3D space (Fig. 8.3 sketches this picture). Individual doublons form a "doublon network." Each doublon is expanded along a 1D line coordinate and bound by two end points, described by gradients of the field, right- and left-moving solitons. Due to the constraint (6.6), the coordinates of different fields have to be synchronized within the junctions of small but finite size η. The constraint (6.6) also dictates that there are no "loose ends"; the body is closed, in itself forming a "universe."

Finishing this lecture let me remind you: In the language of network theory, the junctions of a network are called vertices (in Latin the plural of vertex, a single junction or knot). The connections between vertices, are called "edges". Such a network can be embedded into a 3-dimensional vector space, or higher (but not lower). The doublon network, as we call it in the context of this lecture, does not

Fig. 8.3 Network of six doublons ϕ_α with four vertices, the particles P_i. You may recognize the analogy to "introverted spaces" in Fig. 8.1

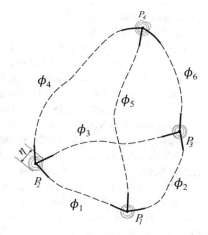

define a vector space by itself! This means that empty spaces inside the meshes of the network are not accessible, as they would in a vector space. They have no physical reality. Only points on the doublons can be attributed by a local coordinate, related to the line distance to a junction or vertex. This is important, since vertices, which are not connected by edges, are "invisible:" there is no physical space between them where light, information, action etc. could be transmitted through. It will be a future task to embed the doublon network into a four-dimensional space-time where the doublons form geodesic trajectories between particles.

A last comment relates to the topology of positive and negative states of energy. The positive states U relate to particles, junctions between the doublons: each particle connects all incoming and outgoing doublons. In contrast, one doublon connects exactly two particles. This defines the network structure, see also above. Furthermore, positive and negative energy states cannot simply annihilate without changing the topology of the whole system: once separated into **MANY** positive and negative states, the system evolves irreversibly.

8.5   Symmetry Breaking in Condensed Matter and Elementary Particle Physics

"Symmetry breaking" is a fundamental concept in materials physics, see the consideration about "phases" and the "order parameter" in Chap. 1. The term "symmetry" here relates to solid materials' crystal symmetry, or the lack of symmetry in the amorphous state, liquid or gas. A particular case is magnetic phase transformation from the ferromagnetic to the paramagnetic phase. In the ferromagnetic state we see spontaneous magnetization of the magnetization $\pm\vec{M}$, which is a vector in 3-dimensional space and can have $+$ or $-$ direction. This is the "symmetry broken state". In contrast, the paramagnetic phase, which is the stable phase above the

Curie temperature T_c. It is called the "symmetric state" with vanishing spontaneous magnetization $\vec{M} = 0$.

In the context of elementary particle physics, the concept of "symmetry breaking" was first introduced by Goldstone [11], referring to the theory of superconductivity, i.e. a phase transformation in condensed matter. Very soon, the publication "Broken Symmetry" by Goldstone et al. [12] appeared, which established the concept of symmetry breaking in elementary particle physics, so-called Goldstone modes. Only two years later, the work of Higgs [13] and Englert and Brout [10] appeared, which was awarded the Nobel prize in 2013 after the so-called "Higgs boson" had been confirmed experimentally [4]. In light of the present concept, the Higgs field, which couples to fermions to give them their mass, can be identified with the phase field as we call it in materials science: An order parameter field in space and time with broken symmetry. The underlying mathematical formalism, Lagrange and Hamiltonian functions are almost identical. The new contribution of the "quantum phase field" is the explicit solution of the minimum energy solution of this field in 1 dimension; the doublon, as detailed. Doublons then are connected to form a doublon network in 3+1 dimensional space-time by the sum constraint of a multi phase-field, Eq. (8.1).

$$\sum_\alpha \phi_\alpha = 1.$$

In conclusion: phase-field theory has two different roots (see Chap. 1): one in thermodynamics and one in wave mechanics. Thermodynamical principles of energy conservation and entropy maximization set the stage. The wave-mechanical picture of phase fields rests on the theory of solitons. Quantum-phase-field theory has roots in de Broglie and Bohm's interpretation of matter as a wave phenomenon; it combines the soliton solution of a non-linear wave equation, the phase-field equation, with a Schrödinger-type linear wave equation in a finite domain. The non-linear wave equation is thus derived from the thermodynamic theory of phase transformation: Ginsburg–Landau theory. We may use other potentials than the double well or double obstacle, as explained earlier. We may use higher gradients in ϕ. The general structure, however, is consistent, having three different contributions, as in every phase-field model: gradient, potential, and bulk (see Chap. 2).

Where is the bulk part in the Hamiltonian (8.6)? The bulk part arises from applying the gradient operator to the wave function! It is defined from the solution of Schroedinger equation in the box formed by the doublon. In the concept, and in "reality" (I think), there is no given "space," no bulk in which objects, "elementary particles," are placed. The elementary particles, which in the phase-field interpretation are junctions or knots between doublons, are defined by gradients of the phase fields. They are the endpoints of the doublons and form the bounds of space. Space and mass are two sides of the same coin. This all may seem "quite philosophical" to the student, or simply "confusing." But it should not discourage you from doing the exercise below [calculating the energy of "space" from (8.10)]. We conclude with the following statements:

- Energy is substantial and conserved.
- There is no fundamental space, no fundamental time.
- Space, a one-dimensional distance, emerges from variation of energy.
- Mass is attributed by positive energy, space by negative energy.
- Two half-sided solitons form the doublon, which is the primitive object of the concept, defining mass and space.
- The doublon belongs to the 3D S(U2) group of spinors. Doublons span the 3D space of our cognition.
- The concept is relativistically invariant and makes predictions at the scale of the universe (see "further reading").

8.6 Exercises

Exercise

Derive the "energy of space" (8.12): $E_\alpha = -\tilde{\alpha} \frac{hc}{48\Omega_\alpha}$ defined by vacuum fluctuation in the 1D box formed by a doublon of size Ω_α.

Further reading

- Hydrodynamic basis of quantum mechanics and its relation to Phase Field: [19]
- Expansion of the universe by repulsive gravitational action on ultra-long distances: [20, 21].

References

1. D. Bohm, A suggested interpretation of the quantum theory in terms of "hidden" variables. I. Phys. Rev. **85**, 166–179 (1952)
2. D. Bohm, A suggested interpretation of the quantum theory in terms of "hidden" variables. II. Phys. Rev. **85**, 180–193 (1952)
3. H. Casimir, On the attraction between two perfectly conducting plates. Proc. Koninklijke Nederlandse Akademie van Wetenschappen **B51**, 793–795 (1948)
4. S. Chatrchyan, et al., Observation of a new boson at a mass of 125 GeV with the CMS experiment at the LHC. Phys. Lett. B **716**(1), 30–61 (2012). ISSN: 0370-2693. https://www.sciencedirect.com/science/article/pii/S0370269312008581
5. L. de Broglie, Nonlinear wave mechanics, in *Trans*, ed. by A.J. Knodel (Elsevier, Amsterdam, 1960)
6. L. de Broglie, L'interpretation de la mechanique ondulatoire par la theorie de la double solution. Proc. Int. School Phys. Enrico Fermi **49**, 346–367 (1971)

7. D.S. DeWitt, Quantum theory of gravity. I. The canonical theory. Phys. Rev. **160**, 1113–1148 (1967)
8. A. Einstein, Die Grundlage der allgemeinen Relativitätstheorie. Ann. Phys. **49**, 769–822 (1916)
9. A. Einstein, Über Gravitationswellen. Sitzungsberichte der Königlich Preussischen Akademie der Wissenschaften Berlin 154–167 (1918)
10. F. Englert, R. Brout, Broken symmetry and the mass of gauge vector mesons. Phys. Rev. Lett. **13**, 321–323 (1964). https://link.aps.org/doi/10.1103/PhysRevLett.13.321
11. J. Goldstone, Field theories with « superconductor » solutions. Il Nuovo Cimento **19**(1), 154–164 (1961). https://doi.org/10.1007/BF02812722
12. J. Goldstone, A. Salam, S. Weinberg, Broken symmetries. Phys. Rev. **127**, 965–970 (1962). https://link.aps.org/doi/10.1103/PhysRev.127.965
13. P.W. Higgs, Broken symmetries and the masses of gauge Bosons. Phys. Rev. Lett. **13**, 508–509 (1964). https://link.aps.org/doi/10.1103/PhysRevLett.13.508
14. L. Hughston, W. Shaw, Classical strings in ten dimensions. Proc. R. Soc. A Math. Phys. Eng. Sci. **414**, 423–431 (1987). https://doi.org/10.1098/rspa.1987.0152
15. T. Kaluza, Zum Unitätsproblem der Physik. Sitzungsberichte der Königlich Preußischen Akademie der Wissenschaften 966–969 (1921)
16. O. Klein, Quantentheorie und fünfdimensionale Relativitätstheorie. Z. Phys. **37**, 895–906 (1926). https://doi.org/10.1007/BF01397481
17. J. Kundin, I. Steinbach, Quantum-phase-field: from the Broglie–Bohm double-solution program to doublon networks. Z. Naturforschung **75**(2a), 155–170 (2020). https://doi.org/10.1515/zna-2019-0343
18. M. Loev, Origin of everything and the 21 dimensions of the universe, in *APS March Meeting Abstracts*. APS Meeting Abstracts, S1.104 (2009)
19. R. Mauri, A non-local phase field model of bohm's quantum potential. Found. Phys. **52**, 52–58 (2001). https://doi.org/10.1007/s10701-021-00454-9
20. I. Steinbach, Quantum-phase-field concept of matter: emergent gravity in the dynamic universe. Z. Naturforschung A **72**(1), (2017). https://doi.org/10.1515/zna-2016-0270
21. I. Steinbach, J. Kundin, F. Varnik, Self similarity of the expanding universe as understood by quantum-phase-fields. arXiv: 2002.12848 [physics.gen-ph] (2020)

Part II
OpenPhase

Chapter 9
Tutorial 1: OpenPhase

9.1 Introduction to the OpenPhase Software Package

OpenPhase is an open-source C++ software project started in 2008 at The Inter-disciplinary Centre for Advanced Materials Simulation, Ruhr-Universität Bochum, Germany. OpenPhase is dedicated to investigating microstructure evolution in materials undergoing first-order phase transformations, capillarity-driven coarsening processes, or a combination of both. Also you may embed the simulation into a macroscopic process simulation which determines boundary conditions on the micro domain, heat fluxed, mechanical loading or others.

OpenPhase is based on the multi-phase-field model as developed by the lecturer and his co-workers. However, the software is not restricted to "Steinbach models." Other models, such as the Khachaturyan scheme of interface homogenization (see Chap. 7), are already implemented to be applied to problems where these models are of advantage, either from a theoretical point of view or simply from a practical point of view for an effective numerical solution. It is straightforward to implement your model in the software. We also strongly encourage scientists all around the world to provide their own special solutions. You are welcome to contribute new modules to the software project!

The commercial version, OP Studio, is meant for users from industry, as well as for users from academia who do not have sufficient experience in numerical simulations to handle the academic code without support. It provides a graphical user interface and additional features like the coupling to thermodynamic databases for multi-component diffusion, as well as finite-strain elasticity and advanced plasticity models. Furthermore, custom solutions for specific user problems can be offered.

The first section of this chapter is dedicated to providing an overview of the OpenPhase structure, including installation, compilation, and executing your first example. In the second section, we explore several representative examples that demonstrate the use of the software and its effectiveness.

© The Author(s) 2023
I. Steinbach, H. Salama, *Lectures on Phase Field*,
https://doi.org/10.1007/978-3-031-21171-3_9

9.1.1 Features

- Open-source library.
- Simplified microstructure creation.
- 1D, 2D and 3D simulations are possible.
- Multiple chemical components, phases, and grains are possible in a multi-phase-field simulation.
- OpenMP and MPI parallelization.
- Tools to extract various statistics data are provided.

9.2 Download

OpenPhase is licensed under the GNU General Public License version 3 (GPLv3) for all open-source applications. We invite you to visit our website at https://openphase.rub.de to obtain your OpenPhase version.

Paid support and commercial licenses for the OpenPhase library can be obtained from the OpenPhase Solutions GmbH. Visit: https://openphase-solutions.com. OpenPhase Solutions GmbH also offers an end-user-oriented standalone phase-field application with extended functionality.

9.3 Installation

To install OpenPhase follow the installation instructions provided here. Make sure to manually install the required packages to ensure a successful installation.

9.3.1 System Requirements

OpenPhase works on **Unix-like** OSs. Therefore, if you would like to use Windows, we advise you to install the Windows Subsystem for Linux (following Microsoft's instructions: https://docs.microsoft.com/en-us/windows/wsl/install#manual-installation-steps) or to install any Linux distribution as a virtual machine on Windows.

Minimum System Requirements

- GNU GCC C++17 compliant compiler (GCC @ 9.0.0 or greater).
- The Fastest Fourier Transform in the West (FFTW) package

9.3.2   Installing the Prerequisites

The following software is required before installing OpenPhase:

- The GCC compiler (g++)
- FFTW library. (www.fftw.org)

The GCC Compiler

1. Open the terminal (Linux: Ctrl+Alt+T)

2. Update your packages using:

```
$ sudo apt-get update
```

3. Install the development package:

```
$ sudo apt install build-essential
```

4. Check g++ compiler version:

```
$ gcc --version
```

FFTW

1. If you are using the `apt` package manager, Installing fftw-dev package is as easy as running the following command on terminal:

   - Non MPI users

```
$ sudo apt-get install fftw-dev libfftw3-3 libfftw3-dev
```

   - MPI users

```
$ sudo apt-get install libfftw3-mpi-dev
```

2. Or you can use the source code instead. Download the newest FFTW package using https://www.fftw.org/download.html.

   - Now navigate to the Downloads folder and extract using:

```
$ tar xvzf FileName.tar.gz
```

   - Navigate to the extracted folder. The installation can be as simple as:

```
$ ./configure
$ make
$ make install
```

   - The FFTW MPI source code can be compiled by following: https://www.fftw.org/fftw3_doc/FFTW-MPI-Installation.html.

9.3.3   OpenPhase (Compilation)

After downloading the source code of OpenPhase as described in Sect. 9.2, extract the source file and navigate to the extracted folder.

Now, compile the source code as follows:

- Non MPI users

```
$ make
```

- MPI users

```
$ make SETTINGS=mpi-parallel
```

This will take a few minutes, depending on your system.

If you got a "**Compilation done**" message on your terminal, you are ready to run your first example.

9.3.4   Update OpenPhase

OpenPhase employs standard versioning, although it is continuously under development with frequent updates. We recommend keeping OpenPhase up to date as you use it to develop your application(s); regular upgrades are advised. New and archived versions of the OpenPhase library can be found on our website at https://openphase.rub.de/download.html.

9.3.5   OpenPhase Application

Structure of an OpenPhase Application

Each application example directory should have at least the following files (Mandatory):

- ProjectInput.opi
- Main.cpp
- Makefile

It may include additional files (Optional):

- Auxiliary input files (e.g. EBSD, Orientations, Geometry, etc.)
- Post Processing files (e.g. Gnuplot, python, etc.)

The structure and parameters for the input file, which has been given the name "ProjectInput.opi", will be discussed in the next section.

Compile Your Application

Compile the application source code as follows:

- Non MPI users

  ```
  $ make
  ```

- MPI users

  ```
  $ make SETTINGS=mpi–parallel
  ```

If the application is functioning properly, the output won't indicate errors. Depending on your compiler version, you may get warnings that can usually be disregarded. Be sure to recompile and test your application after making changes to the source code or updating OpenPhase library.

Run Your Application

- Non MPI users

  ```
  $ ./ApplicationName
  ```

- MPI users

  ```
  $ mpirun –np 4 ApplicationName
  ```

This command tells the computer to run the executable program "Application-Name" on 4 CPUs.

Note: This mpirun command may be useful for novices with simple tasks, but it's not the way to go when you have a large, lengthy application! Additionally, outcomes can vary from machine to machine. When you request 4 processors, for example,

- In the best-case scenario, mpirun detects three additional PCs similar to the one you're using, copies your software to them, and runs it on all four.
- In a less ideal scenario, your current system has four processors, this memory is shared among them, and your program runs;
- In the worst-case scenario, there are less than four processors available, therefore one of them will "timeshare" and alternately run two or more of your process.

9.4   The Input File Structure

OpenPhase is a command-line, non-interactive tool; it anticipates that the simulation input parameters will be provided in the form of an input file. The user must ensure that the input parameters file is placed in the same folder as the simulation example before proceeding. This file is typically referred to as **ProjectInput file**. The OpenPhase software will read the input file and begin the simulation immediately if the data is correct and sufficient. If this is not the case, then OpenPhase will show an error message and expect the user to correct the input file independently.

9.4.1   Step by Step Through the ProjectInput File

This section will walk through a default **ProjectInput file** line by line, explaining exactly what each line does and what it represents.

RunTimeControl

The input file begins with the `RunTimeControl` module input. This controls the simulation time, the frequency of outputs, and whether the user wants to start a new simulation or restart an existing simulation from a specific time step. Figure 9.1 shows all the parameters related to the `RunTimeControl` module. The following gives a description of each parameter:

> **— SimTtl**
> The user can specify a title for the simulation.

```
@RunTimeControl

$SimTtl      Simulation Title                   : Normal grain growth
$nSteps      Number of Time Steps               : 200000
$FTime       Output to disk every (tSteps)      : 100
$STime       Output to screen every (tSteps)    : 100
$LUnits      Units of length                    : m
$TUnits      Units of time                      : s
$MUnits      Units of mass                      : kg
$EUnits      Energy units                       : J
$dt          Initial Time Step                  : 1e-5
$nOMP        Number of OpenMP Threads           : 10
$Restrt      Restart switch (Yes/No)            : No
$tStart      Restart at time step               : 0
$tRstrt      Restart output every (tSteps)      : 10000
```

Fig. 9.1 The `RunTimeControl` section in the `ProjectInput` file

— nSteps

The number of time steps for the entire simulation. To calculate the actual time, you have to multiply this number by Δt.

— FTime, STime

Sets parameters for the frequency of data output to the disk (`FTime`) and to the screen (`STime`).

— LUnits, TUnits, MUnits

All the values are in the SI system of units, which uses the meter, second, and kilogram as base units.

— dt

This is the initial time step (time discretization) Δt. Note that some modules might use an internal time-stepping scheme.

— nOMP

The number of OpenMP threads can be specified if you want to run your simulation in a multithreading mode. The number 1 indicates no active multithreading. Make sure to understand how OpenMP works before running your simulation using multithreading because if you increase it unwisely, it can slow your simulation instead of making it faster.

— Restrt

OpenPhase gives the user the option of either starting a new simulation (No) or restarting from the existing previously saved output (Yes).

— tStart

Suppose you indicate (Yes) in the previous option. In that case, you have to specify the time step from which you want to restart your simulation.

> **— tRstrt**
> Set parameter for the frequency of raw-data output to the disk for restarting.

Settings

The Settings input controls the size and numerical resolution of the simulation domain, and the user must specify the grid spacing and the number of grid points along each dimension. The grid spacing is one of the most critical numerical parameters because it defines the resolution of the simulation. In practice, high resolution means more time, so it is usually necessary to find a middle ground between grid resolution and calculation time. Other information linked to the phases is contained inside the Settings parameters, including the thermodynamic phase of each phase field and various additional information about the phases, such as their aggregate state (solid, liquid, or gas) and the chemical components, among other things, as shown in Fig. 9.2.

> **— Nx, Ny, Nz**
> The simulation system size in the x, y, and z directions given in grid points.

```
@Settings

$Nx            System Size in X Direction                  : 64
$Ny            System Size in Y Direction                  : 64
$Nz            System Size in Z Direction                  : 64
$IWidth        Interface Width (Grid points)               : 5.0
$Resolution    Phase field resolution                      : Single
$dx            Grid Spacing                                : 1e-6

$Phase_0       Name of Phase 0                             : Austenite
$State_0       State of matter of phase 0                  : Solid
$Comp_0        Chemical component of phase 0               : FE
$Nvariants_0   Number of symmetry variants of phase 0      : 1

$Phase_1       Name of Phase 1                             : Martensite
$State_1       State of matter of phase 1                  : Solid
$Comp_1        Chemical component of phase 1               : FE
$Nvariants_1   Number of symmetry variants of phase 1      : 24
```

Fig. 9.2 The Settings section in the ProjectInput file

— **dx**

Sets the grid spacing Δx. It is important to note that this value is the same for all three spatial dimensions.

— **iWidth**

The interface width η of the phase field given in grid points.

— **Phase**

The name of the phase; the running index refers to the thermodynamic phase for a given phase field.

— **State**

The aggregate state of the phase (solid, liquid, or gas).

— **Comp**

The chemical component name.

— **Nvariants**

The number of crystallographic (symmetry/translation/...) variants for a given phase.

BoundaryConditions

The boundary conditions of the simulation domain have to be set individually for every simulations domain side in the ProjectInput file, as shown in Fig. 9.3.

```
@BoundaryConditions

$BC0X          X axis beginning boundary condition      : Periodic
$BCNX          X axis far end boundary condition        : Periodic

$BC0Y          Y axis beginning boundary condition      : Periodic
$BCNY          Y axis far end boundary condition        : Periodic

$BC0Z          Z axis beginning boundary condition      : Periodic
$BCNZ          Z axis far end boundary condition        : Periodic
```

Fig. 9.3 The `BoundaryConditions` section in the `ProjectInput` file

— BCN0X, BCNX, BCN0Y, BCNY, BCN0Z, BCNZ
These keywords define the six individual sides of the calculation domain.

In OpenPhase The following boundary conditions are available:

- **Periodic**: Periodic boundary conditions.
- **NoFlux**: Adiabatic boundary conditions, resulting in zero first-order derivative at the boundary.
- **Free**: Boundary conditions resembling a free surface, resulting in a continuous gradient across the boundary.
- **Fixed**: Fixed (Dirichlet or Neumann) boundary conditions. The value should be set by the user.

Chapter 10
Tutorial 2: OpenPhase Examples

10.1 Normal Grain Growth

Grain growth is a microstructure-evolution phenomenon that occurs in polycrystalline materials after solidification. In this process, some grains are growing at the cost of the shrinkage of other grains or crystallites of a specific orientation. Consequently, the average grain size increases, and the number of grains in the material decreases.

Grain growth is curvature-driven grain-boundary motion between different crystallites with different orientations. The energy of these grain boundaries adds excess free energy to the system. Therefore, the reduction of the grain-boundary energy is the driving force for the movement of the grain boundaries, and the system evolves towards minimizing the total free energy.

As a result, the material's microstructure proceeds to reduce the total interface area. The kinetics of grain boundaries influence the interfacial migration; the energy and mobility of these boundaries control how the grain structure evolves.

10.1.1 Simulation Example

The OpenPhase distribution comes with the `NormalGG` example, which demonstrates the simulation of grain growth. The driving force is curvature, which is the only contribution to the microstructure evolution. Other contributions, such as temperature and concentration, can easily be coupled.

The example shows normal grain growth with isotropic interface energy and mobility, and the user can modify the example to account for the anisotropy of both contributions.

© The Author(s) 2023
I. Steinbach, H. Salama, *Lectures on Phase Field*,
https://doi.org/10.1007/978-3-031-21171-3_10

Table 10.1 Simulation parameters

Parameter	Value	Units
Grid size	101^3	Cells
Grid spacing Δx	1×10^{-6}	m
Time step Δt	1×10^{-5}	s
Interface width η	$5\Delta x$	m
Interface energy σ	0.24	$\frac{J}{m^2}$
Interface mobility μ	1×10^{-7}	$\frac{m^4}{Js}$

The microstructure is generated using Voronoi tessellation. The initial number of grains can be changed directly by the user in the example file. The `ProjectInput.opi` file contains all the input parameters related to the simulation, and the user can modify it for different applications. In this simulation, random crystallographic orientations are assigned to all grains. Periodic boundary conditions are introduced in all directions. The other simulation parameters are listed in Table 10.1.

The reader can find the source code of this example in the OpenPhase distribution in the examples folder. The source code is also included in this chapter (Listing 1). **Warning:** the code presented here is only for illustration and is subject to changes in the future. The reader is therefore referred to the OpenPhase website for the actual documentation.

Step by Step Through the Source Code

Lines 1–23:   Importing different classes from the OpenPhase library that are used in the simulation of this example.

Lines 26–28:   This function sets the initial microstructure as shown in Fig. 10.1b. In this illustrative example, 200 grains have been initialized with the same thermodynamic phase but different phase-field indices based on the Voronoi tessellation technique implemented in the `Initializations` class. The grains have a random crystallographic orientation, indicated by the RGB colors in Fig. 10.1a.

Line 32:   This function sets the interface properties: mobility and energy. In this example, isotropic interface energy and mobility have been used, as shown in Fig. 10.2. However, the user can use different anisotropy model forces for the interface energy and mobility in the `ProjectInput` file.

Line 33:   This function calculates the interface-curvature-related driving force. It includes the multi-junction energy as indicated in Eq. (6.1).

Line 34:   This function limits the phase-field increments for all present phase-field pairs so that the actual phase-field values are within their natural limits of 0 and 1.

Line 35:   This function merges the increments into the phase fields. It finalizes the phase-field calculations by setting the boundary conditions, calculating the

```
1    #include "Settings.h"
2    #include "RunTimeControl.h"
3    #include "DoubleObstacle.h"
4    #include "PhaseField.h"
5    #include "Initializations.h"
6    #include "BoundaryConditions.h"
7    #include "InterfaceProperties.h"
8    #include "Tools/TimeInfo.h"
9    #include "Info.h"
10
11   using namespace std;
12   using namespace openphase;
13
14   int main()
15   {
16       string InputFile = "ProjectInput.opi";
17       Settings                 OPSettings;
18       RunTimeControl           RTC(OPSettings);
19       PhaseField               Phi(OPSettings);
20       DoubleObstacle           DO(OPSettings);
21       InterfaceProperties      IP(OPSettings);
22       BoundaryConditions       BC(OPSettings);
23       TimeInfo                 Timer(OPSettings, "Execution Time
       ↪  Statistics");
24
25       //generating initial grain structure using Voronoi algorithm
26       int number_of_grains = 200;
27       size_t GrainsPhase = 0;
28       Initializations::VoronoiTesselation(Phi, BC, OPSettings,
       ↪  number_of_grains, GrainsPhase);
29
30       for(RTC.tStep = RTC.tStart; RTC.tStep <= RTC.nSteps;
       ↪  RTC.IncrementTimeStep())
31       {
32           IP.Set(Phi);
33           DO.CalculatePhaseFieldIncrements(Phi, IP);
34           Phi.NormalizeIncrements(BC, RTC.dt);
35           Phi.MergeIncrements(BC, RTC.dt);
36
37           if (RTC.WriteVTK())
38           {
39               Phi.WriteVTK(RTC.tStep, OPSettings);
40           }
41           if (RTC.WriteToScreen())
42           {
43               double I_En = DO.AverageEnergyDensity(Phi, IP);
44               std::string message = Info::GetStandard("Interface energy
               ↪  density", to_string(I_En));
45               Info::WriteTimeStep(RTC, message);
46               Timer.PrintWallClockSummary();
47           }
48       }
49       return 0;
50   }
```

Listing 1: Source code for the NormalGG example in OpenPhase

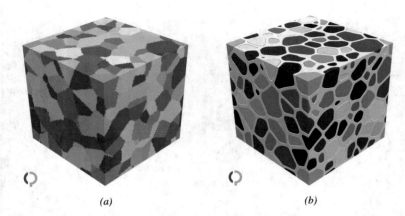

(a)   *(b)*

Fig. 10.1 Initial microstructure, displayed in different modes: (**a**) RGB colors; (**b**) as phase-field index, where the diffuse interface region is displayed in light grey

```
@InterfaceProperties

$EnergyModel_0_0     Interface energy model          : ISO
$Sigma_0_0           Interface energy                : 0.24

$MobilityModel_0_0   Interface mobility model        : ISO
$Mu_0_0              Interface mobility              : 1.0e-7
```

Fig. 10.2 The InterfaceProperties module in the ProjectInput file

derivatives, setting the flags that mark interfaces, collecting the volume for each phase field, and calculating the thermodynamic phase fractions.

Line 39: This function writes an output file in VTK format, which contains the information of the phase field (such as the interfaces, the junctions, and the phase fractions of the different phases). *III ParaView* can be used to visualize the VTK files, and the output results can be seen in Fig. 10.3.

Lines 41–47: This code block prints different simulation information on the screen.

Results

Figure 10.3 shows the microstructural evolution for the given parameters, in which the grain boundaries appear as grey regions separating individual grains.

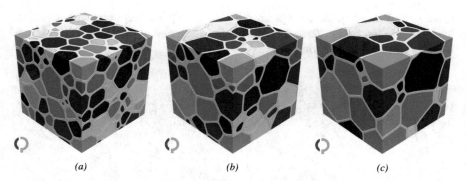

Fig. 10.3 Evolution of the microstructure at different times t (in time steps). (**a**) $t = 500$. (**b**) $t = 1500$. (**c**) $t = 3000$

10.2 Dendritic Solidification

Phase-field simulation can be used to model the solidification of metallic alloys. Interface velocities are well represented by a Gibbs–Thomson relation. This is known as the Stefan problem when linked to heat and solute redistribution at the interface and long-range transport in bulk phases. Interfacial energy anisotropy is weak and can be considered a perturbation of a spherical Wulff shape. Even though the solid and liquid have different densities, no significant stresses are developed at the interface during growth. In comparison to solute diffusion in the melt, heat conduction in both phases is almost the same, and solute diffusion in the solid can be ignored [1]. In general, the evolution of the dendrite interface area is characterized by a rapid increase during the growth of side branches, and this is followed by a decrease due to capillarity and coalescence of adjacent structures.

10.2.1 Simulation Example

The OpenPhase distribution provides the `SolidificationFeC` example, which is a simulation of dendritic solidification.[1] The code presented here is written considering a 2D case. However, it can be modified to account for the 3D case. In this example, two different simulations have been considered: 2D equiaxed dendritic solidification for a single grain and for multiple grains with different orientations growing from the melt.

The reader can find the source code for this example in the OpenPhase distribution in the examples folder. The source code is also included in the chapter

---

[1] The code presented here was developed in collaboration by a Chinese exchange student, MSc. Jianghai Cao from Chongqing University, China, during his interim at RUB.

(Listing 2). **Warning:** the code presented here is only for illustration and is subject to changes in the future. Therefore, the reader is referred to the OpenPhase website for the actual documentation.

```
1    #include "Settings.h"
2    #include "RunTimeControl.h"
3    #include "InterfaceProperties.h"
4    #include "DoubleObstacle.h"
5    #include "PhaseField.h"
6    #include "DrivingForce.h"
7    #include "Composition.h"
8    #include "Temperature.h"
9    #include "EquilibriumPartitionDiffusionBinary.h"
10   #include "BoundaryConditions.h"
11   #include "Initializations.h"
12   #include "Tools/TimeInfo.h"
13   #include "Tools/MicrostructureAnalysis.h"
14   using namespace std;
15   using namespace openphase;
16   /*********** <<< The Main >>> ***********/
17   int main(int argc, char *argv[])
18   {
19       Settings                               OPSettings;
20       OPSettings.ReadInput();
21       RunTimeControl                         RTC(OPSettings);
22       PhaseField                             Phi(OPSettings);
23       DoubleObstacle                         DO(OPSettings);
24       InterfaceProperties                    IP(OPSettings);
25       EquilibriumPartitionDiffusionBinary    DF(OPSettings);
26       Composition                            Cx(OPSettings);
27       Temperature                            Tx(OPSettings);
28       DrivingForce                           dG(OPSettings);
29       BoundaryConditions                     BC(OPSettings);
30       TimeInfo                               Timer(OPSettings, "Execution Time
31    ↪     Statistics");
32       int idx0 = Initializations::Single(Phi, 0, BC, OPSettings);
33       Phi.FieldsStatistics[idx0].State = AggregateStates::Liquid;
34       int idx1 = Phi.PlantGrainNucleus(1, OPSettings.Nx/2, OPSettings.Ny/2,
     ↪     OPSettings.Nz/2);
35       Phi.FieldsStatistics[idx1].State = AggregateStates::Solid;
36
37       Cx.SetInitial(BC, Phi);
38       Tx.SetInitial(BC);
39       DF.SetDiffusionCoefficients(Phi, Tx);
40
41       //-------------- The Time Loop -------------//
42       for(RTC.tStep = RTC.tStart; RTC.tStep <= RTC.nSteps;
     ↪     RTC.IncrementTimeStep())
43       {
44           IP.Set(Phi);
45           DF.CalculateInterfaceMobility(Phi, Cx, Tx, BC, IP);
46           DO.CalculatePhaseFieldIncrements(Phi, IP);
47           DF.GetDrivingForce(Phi, Cx, Tx, dG);
48           dG.Average(Phi, BC);
49
50           dG.MergePhaseFieldIncrements(Phi, IP);
51           Phi.NormalizeIncrements(BC, RTC.dt);
52           DF.Solve(Phi, Cx, Tx, BC, RTC.dt);
53           Tx.Set(BC, Phi, RTC.dt);
54           Phi.MergeIncrements(BC, RTC.dt);
```

```
55            //  Output to file
56            if (RTC.WriteVTK())
57            {
58                // Write data in VTK format
59                Phi.WriteVTK(RTC.tStep,OPSettings);
60                Cx.WriteVTK(RTC.tStep,OPSettings);
61                Tx.WriteVTK(RTC.tStep,OPSettings);
62                IP.WriteVTK(Phi,RTC.tStep);
63            }
64            // Output to screen
65            if(RTC.WriteToScreen())
66            {
67                double I_En = DO.AverageEnergyDensity(Phi, IP);
68                std::string message = Info::GetStandard("Interface energy
                 ↪  density", to_string(I_En));
69                Info::WriteTimeStep(RTC, message);
70                dG.PrintDiagnostics();
71                Phi.PrintVolumeFractions();
72                Timer.PrintWallClockSummary();
73            }
74        } //end time loop
75        return 0;
76    }
```

Listing 2: Source code for the `SolidificationFeC` example in OpenPhase

Step by Step Through the Source Code
1: Equiaxed Dendritic Solidification for a Single Grain
Lines 1–30:   Importing different classes from the OpenPhase library that are used in the simulation of this example.
Lines 32–35:   These functions set the initial microstructure as illustrated in Fig. 10.4a. There is a matrix with a single phase representing the liquid, which has a phase index equal to 0, and we plant a grain nucleus in the middle of our simulation domain, which represents the solid phase and has a phase index equal to 1.
Lines 37–38:   These functions set the initial composition, and the temperature and diffusion coefficients for the system. They take their inputs from the `ProjectInput` file.
Line 44:   This function sets the interface properties: mobility and energy. In this example, anisotropic driving forces for the interface energy and mobility have been set in the `ProjectInput` file.
Line 45:   This function calculates concentration-dependent mobility.
Line 46:   This function calculates the interface-curvature-related driving force. It includes the triple-junction energy as indicated in Eq. (6.1).
Line 47:   This function calculates the driving force for each point.
Line 48:   This function averages the driving forces across the interface.

Line 50: This function calculates a time-independent phase-field increment by converting the local driving forces. An additional noise term can be applied, as can a stability mechanism to strengthen the numerical phase-field profile.

Line 51: This function limits phase-field increments for all present phase-field pairs so that the actual phase-field values are within their natural limits of 0 and 1.

Line 52: This function calculates the change of total concentrations in one time step taking into account cross terms.

Line 53: This function sets the temperature according to the cooling rate; it applies equal temperature increments to all points.

Line 54: This function merges the increments into the phase fields. It finalizes the phase-field calculations by setting the boundary conditions, calculating the derivatives, setting the flags that mark interfaces, collecting the volume of each phase field, and calculating the thermodynamic phase fractions.

Lines 57–65: This code block writes different output files in VTK format, which contains the information of the phase field (such as the interfaces, the junctions, and the phase fractions of the different phases), composition, temperature, and interface properties. *III ParaView* can be used to visualize the VTK files.

Lines 66–74: This code block prints different simulation information on the screen.

2: Equiaxed Dendritic Solidification for Multiple Grains with Different Orientations

For this case, the code stays the same as discussed above with a little modification to the initialization part, in which line 34 can be replaced by the following line:

This function plants random nucleation sites in the simulation domain with different orientations. The number of particles can be chosen by replacing `NoOfParticles` with an integer number.

Results

Figures 10.4 and 10.5 show that side branches emerge as the dendritic tips extend into the undercooled melt, resulting in a complicated solid–liquid interface morphology. Once you reach a certain distance behind the dendritic tip, a slower evolution begins to take place, which occurs at or around the point of phase equilibrium. Capillarity and delayed solidification play a significant role in determining the interface dynamics at this stage.

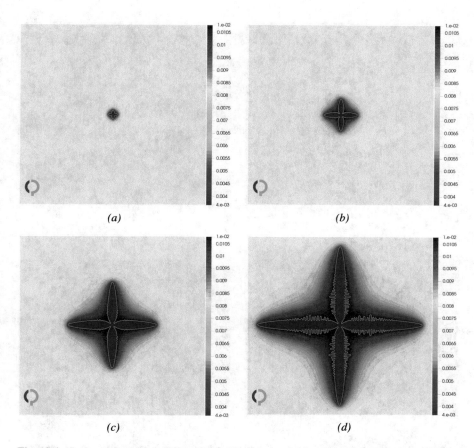

Fig. 10.4 Equiaxed dendritic solidification in 2D. (**a**) $t = 0.05$ s. (**b**) $t = 0.25$ s. (**c**) $t = 0.75$ s. (**d**) $t = 1.5$ s

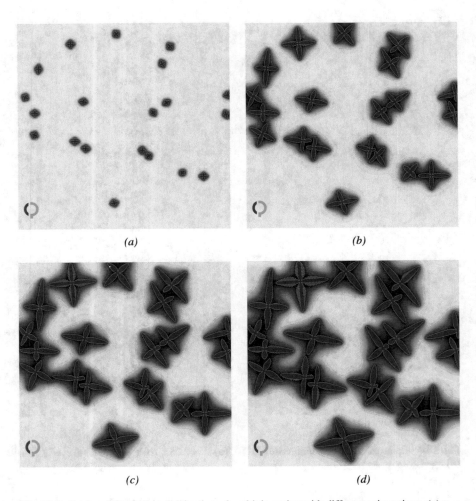

Fig. 10.5 Equiaxed dendritic solidification of multiple grains with different orientations. (**a**) $t =$ 0.05 s. (**b**) $t = 0.25$ s. (**c**) $t = 0.75$ s. (**d**) $t = 1.5$ s

Reference

1. I. Steinbach, Phase-field models in materials science. Modell. Simul. Mater. Sci. Eng. **17**, 073001 (2009)

Appendix A

A.1 Simple Dendrite Code

This section presents the "100-line C++ code" that we use to teach the students how to simulate a dendritic growth similar to the example illustrated in Chap. 1. A good exercise for the reader is to choose suitable parameters and correctly repeat the simulation given in "Example—Dendritic Growth" in Chap. 1.

```cpp
 1  #include <cmath>
 2  #include <fstream>
 3  #include <sstream>
 4  #include <iostream>
 5  #include <cstdlib>
 6  #include <complex>
 7  #include <random>
 8
 9  using namespace std;
10
11  // Discretization parameters
12  const int    Nx      = 300;        // Discretization x-direction
13  const int    Ny      = 300;        // Discretization y-direction
14  const int    Nt      = 1000000000; // Number of time steps
15  const int    BCells  = 1;          // Number of boundary cells
16  double       dx      = 0.03;       // Grid spacing in x-direction [m]
17  double       dy      = 0.03;       // Grid spacing in y-direction [m]
18  double       dt      = 2.0e-4;     // Size of time step [s]
19
20  // Kobayashi's parameters (not exactly his..)
21  double epsilon   = 0.010;      // Gradient energy coefficient
22  double tau       = 3.0e-4;     // Inverse of interface mobility [s]
23  double alpha     = 0.7;        // Coefficient of driving force
24  double Gamma     = 10.0;       // Coefficient of driving force
25  double delta     = 0.04;       // Anisotropy in (0,1)
26  double theta0    = M_PI/2;     // Rotation of the crystal orientation
27  int    sym       = 8;          // Number of symmetry fold
28  double K         = 2.0;        // Referrers to latent heat
         ↪ (no-dimension)
29  double T0        = 0.0;        // Initial temperature
30  double Te        = 1.0;        // Equilibrium temperature
         ↪ (no-dimension)
31  int    RandomSeed = 125;       // Defines the set of random numbers
32  double ampl      = 0.02;       // Amplitude of noise
```

© The Author(s) 2023
I. Steinbach, H. Salama, *Lectures on Phase Field*,
https://doi.org/10.1007/978-3-031-21171-3

```
33
34      // Misc parameters
35      int     tOut            = 100;       // Output distance in time steps
36      double  t               = 0.0;       // Simulation start time [s]
37      double  Radius          = 0.3;       // Initial radius of spherical grain
38      double  PhiPrecision    = 1.e-9;     // Phase-field cut off
39      double  TempPrecision   = 1.e-3;     // Precision of Gauss-Seidel algorithm
40      int     TempMaxIter     = 10000;     // Maximum number of Gauss-Seidel
        ↪    iterations
41
42
43
44      // Phase-field and Temperature storage
45      const int Mx = Nx + 2 * BCells;      // Shorthand for entire storage loop
46      const int My = Ny + 2 * BCells;      // Shorthand for entire storage loop
47
48      double Phi      [Mx][My];            // Phase-field storage
49      double PhiDot   [Mx][My];            // dPhi_dt storage
50      double Temp     [Mx][My];            // Temperature storage
51      double Eps      [Mx][My];            // Laplace parameter
52      double EpsP     [Mx][My];            // Derivative of Laplace parameter
53      double Noise    [Mx][My];            // Storage for noise
54
55
56      /** Calculates Inclination angel
57       * \param    i       position on grid in x-direction
58       * \param    j       position on grid in y-direction
59       * \return           angel theta in [0,pi]
60       **/
61      double CalculateTheta(int i, int j)
62      {
63          double dPhiX    = (Phi[i+1][j  ] - Phi[i-1][j  ])/(2*dx);
64          double dPhiY    = (Phi[i  ][j+1] - Phi[i  ][j-1])/(2*dy);
65          double dPhiNorm = sqrt( pow(dPhiX, 2) + pow(dPhiY, 2));
66
67          double theta = 0.0;
68          if (dPhiNorm > 1.e-10)
69          {
70              double NormX  = dPhiX/dPhiNorm;
71              double NormY  = dPhiY/dPhiNorm;
72              theta += atan2(NormY, NormX);
73          }
74          return theta;
75      };
76
77      /** Anisotropy function
78       * \param    theta   Angel between x-axis and interface normal
79       * \return           Amplitude of Laplace term
80       **/
81      double sigma(double theta)
82      {
83          double value = 1 + delta * cos( sym * (theta - theta0));
84          return value;
85      };
86
87      /** First derivative of anisotropy function
88       * \param    theta   Angel between x-axis and interface normal
89       * \return           Amplitude of Laplace term
90       **/
91      double sigmaP(double theta)
92      {
93          double value = - delta * sym * sin( sym * (theta - theta0));
94          return value;
95      };
96
97      /** Initializes a grain of a certain radius in a supercooled melt
98       * \param    Radius  Radius of grain
```

```
99      **/
100     void InitializeSupercooledGrain(double Radius)
101     {
102         // Initialization
103         const double   x0 = Nx/2 * dx;
104         const double   y0 = 0;
105         for (int i = 0; i < Nx + 2 * BCells; i++)
106         for (int j = 0; j < Ny + 2 * BCells; j++)
107         {
108             // Initialize the phase Field
109             double r = sqrt(pow(i*dx-x0,2) + pow(j*dy-y0,2));
110             if ( r < Radius)
111             {
112                 Phi        [i][j] = 1.0;
113                 PhiDot     [i][j] = 0.0;
114             }
115             else
116             {
117                 Phi        [i][j] = 0.0;
118                 PhiDot     [i][j] = 0.0;
119             }
120             // Initialize temperature
121             Temp   [i][j] = T0;
122         }
123     }
124
125     /** Calculate Laplacian of the phase-field
126      * \param      i       Position on grid in x-direction
127      * \param      j       Position on grid in y-direction
128      * \return             Laplacian of phase-field
129      **/
130     double LaplacePhi(int i, int j)
131     {
132         double Laplacian = 0.0;
133         Laplacian += 2./3.*(Phi[i+1][j  ] - 2.0*Phi[i][j] + Phi[i-1][j
             ↪  ])/(dx*dx);
134         Laplacian += 2./3.*(Phi[i  ][j+1] - 2.0*Phi[i][j] + Phi[i
             ↪  ][j-1])/(dy*dy);
135         Laplacian += 1./3.*(Phi[i+1][j+1] + Phi[i+1][j-1] + Phi[i-1][j+1] +
             ↪  Phi[i-1][j-1] - 4*Phi[i][j])/(pow(sqrt(2.)*dx,2));
136         return Laplacian;
137     }
138
139     /** Calculate Laplacian of the temperature
140      * \param      i       Position on grid in x-direction
141      * \param      j       Position on grid in y-direction
142      * \return             Laplacian of temperature
143      **/
144     double LaplaceTemp(int i, int j)
145     {
146         double Laplacian = 0.0;
147         Laplacian += 2./3.*(Temp[i+1][j  ] - 2.0*Temp[i][j] + Temp[i-1][j
             ↪  ])/(dx*dx);
148         Laplacian += 2./3.*(Temp[i  ][j+1] - 2.0*Temp[i][j] + Temp[i
             ↪  ][j-1])/(dy*dy);
149         Laplacian += 1./3.*(Temp[i+1][j+1] + Temp[i+1][j-1] + Temp[i-1][j+1] +
             ↪  Temp[i-1][j-1] - 4*Temp[i][j])/(pow(sqrt(2.)*dx,2));
150         return Laplacian;
151     }
152
153     /** Writes output to file
154      * \param      tStep     time step
155      **/
156     void WriteToFile(int tStep)
157     {
158         {
159             stringstream outbufer;
```

```
160        outbufer << "# vtk DataFile Version 3.0\n";
161        outbufer << "PhaseField\n";
162        outbufer << "ASCII\n";
163        outbufer << "DATASET STRUCTURED_GRID\n";
164        outbufer << "DIMENSIONS " << Nx << " " << Ny << " " << 1 << "\n";
165        outbufer << "POINTS " <<  Nx*Ny*1 << " int\n";
166
167        for(int j = 1; j < Nx+1; ++j)
168        for(int i = 1; i < Ny+1; ++i)
169        {
170            outbufer << i << " " << j << " " << 1 << "\n";
171        }
172        outbufer << "\n";
173        outbufer << "POINT_DATA " << Nx*Ny*1 << "\n";
174
175        outbufer << "SCALARS PhaseFields double 1" << " \n";
176        outbufer << "LOOKUP_TABLE default" << " \n";
177        for(int j = 1; j < Nx+1; ++j)
178        for(int i = 1; i < Ny+1; ++i)
179        {
180            outbufer << double(Phi[i][j]) << " \n";
181        }
182
183        outbufer << "SCALARS Temperature double 1" << " \n";
184        outbufer << "LOOKUP_TABLE default" << " \n";
185        for(int j = 1; j < Nx+1; ++j)
186        for(int i = 1; i < Ny+1; ++i)
187        {
188            outbufer << double(Temp[i][j]) << " \n";
189        }
190
191        stringstream filename;
192        filename << "PhaseField_" << tStep << ".vtk";
193        string FileName = filename.str();
194
195        ofstream vtk_file(FileName.c_str());
196        vtk_file << outbufer.rdbuf();
197        vtk_file.close();
198    }
199 }
200
201 /** Main Kobayashi code **/
202 int main()
203 {
204     // initialize random seed
205     srand(RandomSeed);
206     InitializeSupercooledGrain(Radius);
207     WriteToFile(0);
208
209     // Main time loop
210     for (int tStep = 1; tStep < Nt; tStep++)
211     {
212         // Calculate gradient coefficient
213         for (int i = BCells; i < Nx + BCells; i++)
214         for (int j = BCells; j < Ny + BCells; j++)
215         {
216             if ((Phi[i][j] > PhiPrecision) and (Phi[i][j] < 1. -
                 ↪ PhiPrecision))
217             {
218                 double theta = CalculateTheta(i,j);
219                 Eps [i][j] = epsilon * sigma(theta);
220                 EpsP[i][j] = epsilon * sigmaP(theta);
221             }
222             else
223             {
224                 Eps [i][j] = epsilon;
225                 EpsP[i][j] = 0.0;
```

```
226                          }
227                          double noise = Phi[i][j] * (1 -Phi[i][j]) * ampl;
228                          noise *= Noise[i][j];
229                          Eps[i][j] += noise;
230                      }
231
232              // Calculate PhiDot
233              for (int i = BCells; i < Nx + BCells; i++)
234              for (int j = BCells; j < Ny + BCells; j++)
235              {
236                  PhiDot[i][j] = 0.0;
237                  // Calculate driving force m
238                  if ((Phi[i][j] > PhiPrecision) and (Phi[i][j] < 1. -
                    ↪ PhiPrecision))
239                  {
240                      double m      = (alpha/M_PI) * atan(Gamma*(Te -
                        ↪ Temp[i][j]));
241                      PhiDot[i][j] += Phi[i][j]*(1.0 - Phi[i][j])*(Phi[i][j] -
                        ↪ 0.5 + m);
242                  }
243
244                  // Calculate Laplacian-term
245                  PhiDot[i][j] -= (Eps[i+1][j] * EpsP[i+1][j] * (Phi[i+1][j+1] -
                    ↪ Phi[i+1][j-1])
246                                      - Eps[i-1][j] * EpsP[i-1][j] * (Phi[i-1][j+1] -
                                      ↪ Phi[i-1][j-1]))/(4*dx*dy);
247                  PhiDot[i][j] += (Eps[i][j+1] * EpsP[i][j+1] * (Phi[i+1][j+1] -
                    ↪ Phi[i-1][j+1])
248                                      - Eps[i][j-1] * EpsP[i][j-1] * (Phi[i+1][j-1] -
                                      ↪ Phi[i-1][j-1]))/(4*dx*dy);
249                  double dPhiX = (Phi[i+1][j] - Phi[i-1][j])/(2*dx);
250                  double dPhiY = (Phi[i][j+1] - Phi[i][j-1])/(2*dy);
251                  double dEpsX = (Eps[i+1][j] - Eps[i-1][j])/(2*dx);
252                  double dEpsY = (Eps[i][j+1] - Eps[i][j-1])/(2*dy);
253                  PhiDot[i][j] += 2*Eps[i][j]*(dPhiX*dEpsX + dPhiY*dEpsY);
254                  PhiDot[i][j] += pow(Eps[i][j],2) * LaplacePhi(i,j);
255                  PhiDot[i][j] /= tau;
256              }
257
258              // Update Phase-field
259              for (int i = BCells; i < Nx + BCells; i++)
260              for (int j = BCells; j < Ny + BCells; j++)
261              {
262                  if (PhiDot[i][j] != 0.0)
263                  {
264                      // Update phase field
265                      Phi   [i][j] += dt * PhiDot[i][j];
266
267                      // Limit phase field
268                      if      (Phi[i][j] <     PhiPrecision) Phi[i][j] = 0.0;
269                      else if (Phi[i][j] > 1 - PhiPrecision) Phi[i][j] = 1.0;
270                  }
271              }
272
273              // Apply adiabatic boundary conditions
274              for (int i = 0; i < Nx + 2*BCells; i++)
275              {
276                  Phi[i][0 ] = Phi[i][1  ];
277                  Phi[i][Ny] = Phi[i][Ny-1];
278              }
279              for (int j = 0; j < Ny + 2*BCells; j++)
280              {
281                  Phi[0 ][j] = Phi[1  ][j];
282                  Phi[Nx][j] = Phi[Nx-1][j];
283              }
284
285              // Calculate TempDot (implicit scheme with Gauss-Seidel)
```

```
286        double MaxDTemp    = 0.0; // Maximum of Temp change in on iteration
287        int     iterations = 0;   // Iteration counter
288        do
289        {
290            MaxDTemp = 0.0;
291            // Save current Temperature in TOld
292            for (int i = BCells; i < Nx + BCells; i++)
293            for (int j = BCells; j < Ny + BCells; j++)
294            {
295                double TempOld = Temp[i][j];
296                double aij = (- 2.*dt/pow(dx,2) - 2.*dt/pow(dy,2) - 1);
297                double bij = - K * PhiDot[i][j] * dt - Temp[i][j];
298                double neightbours = 0.0;
299                neightbours += dt/pow(dx,2) * Temp[i-1][j  ];
300                neightbours += dt/pow(dy,2) * Temp[i  ][j-1];
301                neightbours += dt/pow(dx,2) * Temp[i+1][j  ];
302                neightbours += dt/pow(dy,2) * Temp[i  ][j+1];
303
304                Temp[i][j] = 1./aij * ( bij - neightbours );
305
306                MaxDTemp = max(MaxDTemp, abs( Temp[i][j] - TempOld));
307
308                iterations++;
309            }
310            // Apply adiabatic boundary conditions
311            for (int i = 0; i < Nx + 2*BCells; i++)
312            {
313                Temp[i][0  ] = Temp[i][1  ];
314                Temp[i][Ny ] = Temp[i][Ny-1];
315            }
316            for (int j = 0; j < Ny + 2*BCells; j++)
317            {
318                Temp[0  ][j] = Temp[1  ][j];
319                Temp[Nx ][j] = Temp[Nx-1][j];
320            }
321        }
322        while ((MaxDTemp > TempPrecision) and (iterations < TempMaxIter));
323
324        // Calculate calculation time of time step
325        t += dt; //increase the time
326        // Write into VTK File
327        if( tStep%tOut == 0)
328        {
329            WriteToFile(tStep);
330            cout << "Time step: " << tStep << " (Simulaton Time: " << t <<
               ↪   " s)" << endl;
331        }
332
333    } // end of time loop
334    return 0;
335 }
```

Index

© The Author(s) 2023
I. Steinbach, H. Salama, *Lectures on Phase Field*,
https://doi.org/10.1007/978-3-031-21171-3

Printed in the United States
by Baker & Taylor Publisher Services